Small Bodies
of the
Solar System
A Guided Tour for Non-Scientists

Small Bodies

of the

Solar System

A Guided Tour for Non-Scientists

Hans Rickman

PAS Space Research Centre, Poland
Uppsala University, Sweden

We World Scientific

NEW JERSEY · LONDON · SINGAPORE · BEIJING · SHANGHAI · HONG KONG · TAIPEI · CHENNAI · TOKYO

Published by

World Scientific Publishing Europe Ltd.

57 Shelton Street, Covent Garden, London WC2H 9HE

Head office: 5 Toh Tuck Link, Singapore 596224

USA office: 27 Warren Street, Suite 401-402, Hackensack, NJ 07601

Library of Congress Cataloging-in-Publication Data
Names: Rickman, H. (Hans), author.
Title: Small bodies of the solar system : a guided tour for non-scientists /
    Hans Rickman, PAS Space Research Centre, Poland, Uppsala University, Sweden.
Description: New Jersey : World Scientific, [2022] | Includes bibliographical references and index.
Identifiers: LCCN 2021016036 (print) | LCCN 2021016037 (ebook) |
    ISBN 9781800610514 (hardcover) | ISBN 9781800610606 (paperback) |
    ISBN 9781800610521 (ebook for institutions) | ISBN 9781800610538 (ebook for individuals)
Subjects: LCSH: Solar system--Popular works.
Classification: LCC QB501.2 .R53 2022 (print) | LCC QB501.2 (ebook) | DDC 523.2--dc23
LC record available at https://lccn.loc.gov/2021016036
LC ebook record available at https://lccn.loc.gov/2021016037

British Library Cataloguing-in-Publication Data
A catalogue record for this book is available from the British Library.

For any available supplementary material, please visit
https://www.worldscientific.com/worldscibooks/10.1142/Q0310#t=suppl

Desk Editors: Jayanthi Muthuswamy/Michael Beale/Shi Ying Koe

Typeset by Stallion Press
Email: enquiries@stallionpress.com

This book is dedicated to the memory of my wife,
Bożenna Rickman

Preface

What is the solar system? Ask anyone, and you'll probably get the answer that the solar system is the Sun and the planets moving in orbits around the Sun. The Earth is one of these planets. If the person who answers your question is interested in astronomy, you may also learn that there is also a lot of small rubbish in the solar system taking the form of asteroids and comets.

There is nothing incorrect about calling asteroids and comets with a very small total mass "rubbish". If the small bodies are ignored, one neglects only a constituent of small total mass. The observed small bodies weigh no more than 1/100th of the Earth's mass altogether, and our Earth only contributes about 1/400th of the total mass of the planets. Why should one take an interest in something so infinitesimal?

This is, of course, a relevant question, but one then forgets that the small bodies carry with them the history of the solar system all the way back to its inception. Without the small bodies of the solar system, we would fumble in the dark to understand what happened when the solar system was young. Moreover, these small bodies have an entirely different relationship with the Earth and its inhabitants than the major planets.

This book has been written to explain why small bodies of the solar system are interesting and so much research using spacecraft, telescopes and computers is spent in order to understand them better. I promise that the answer will be convincing!

About the Author

Hans Rickman is a Professor at the Space Research Centre of the Polish Academy of Sciences and Professor Emeritus of Uppsala University, Sweden. He is a member of the Royal Swedish Academy of Sciences and the International Academy of Astronautics. He has previously held positions as General Secretary of the International Astronomical Union (2000–2003), President of the Swedish Astronomical Society (1996–2000) and Associate Editor of *Earth, Moon & Planets* (1993–2014). In recognition of his contribution to planetary and solar system sciences, Rickman was awarded the David Bates Medal of the European Geosciences Union in 2012. Rickman has had an asteroid named after him (Asteroid 3692) and is author of *Origin and Evolution of Comets* (World Scientific, 2017).

Acknowledgements

My heartfelt thanks go to Adam Juchniewicz for helping to improve the English language of this book and to Tomasz Wiśniowski for producing Fig. 4.2. I am grateful to the following colleagues for providing me with illustrations: Horst Uwe Keller (Fig. 1.2), Julio Fernández (Fig. 2.2), Birger Schmitz (Fig. 3.5), Martin Bizzarro and Zita Fihl (Fig. 3.7), Jürgen Blum (Fig. 6.2), Kleomenis Tsiganis (Fig. 6.3) and Kevin Zahnle (Fig. 7.4).

Contents

Chapter 1

What Is a Comet? Chasing the Nucleus

It was the spring of 1976. I was spending a night at the Stockholm Observatory at Saltsjöbaden where I pursued my doctoral studies. My thesis concerned comets, and an unusually bright comet had been discovered by the Danish astronomer Richard West. The night was cold and clear and, before the break of dawn, the comet would rise above the horizon. I was going to take a photograph of it with one of the Observatory's telescopes.

I did not know exactly what West's comet would look like, and only knew that it had brightened rapidly as it rounded the Sun. I felt a great excitement, and I was nervous and expectant before my appointment with the comet. I did not sleep at all that night; I simply waited. Now and then I went outside to check that the clear weather persisted. Sure enough, each time I went out I saw the beautiful starry sky.

About an hour before the rise of the comet, I scouted out across Baggen bay toward the horizon in the East and saw something that caused me great distress. I thought I saw clouds streaking out, standing out with a dim light against the black heaven. Did it have to cloud up right now? Was my wait in vain? But then I realised what was happening. It wasn't clouds, but an enormous comet tail that I was witnessing!

I was so fascinated that I nearly forgot to take any pictures. As time progressed, I saw more and more of the comet tail, and finally, the brightly shining comet head rose above the horizon. The appearance of the whole comet in the first light of dawn over the bays

1

Fig. 1.1. Comet West (C/1975 V1) at the time of its peak brightness in March 1976. Credit: J. Linder/ESO. License: Creative Commons Attribution 4.0 International.

and isles of the archipelago was truly overwhelming, almost eerie, and I shall never forget it (see Fig. 1.1).

Yet in hindsight, I feel a little ashamed. I had not realised that a comet rising late at night before the Sun causes the tail to lead ahead. Indeed, the comet's tail always points away from the Sun. But my recollection of this memory leads to my next question, what *are* comets?

1.1. Activity and Brightness

In the very centre of the shining haze that constitutes the comet head, there is a solid little celestial body, which we call the comet nucleus. The word is an excellent match since this nucleus incorporates almost all the material of the comet and is its only lasting property.

The rest — that which shines brightly enough to see — is merely a temporary secretion from the nucleus on its way to final dispersal into cosmic space. This phenomenon may be likened to New Year's fireworks sent out by the nucleus each time it passes closest to the Sun and a new orbit commences.

Comets all behave very differently — though generally they all follow the rule of reaching maximum brightness when the distance to the Sun is smallest. Only very few reach the same brightness as comet West, but astronomers can always observe other, more modest comets with their telescopes. Why is this? What is it that dictates if a comet becomes the king of the sky or remains unknown and unimportant?

There are two answers to this question, and both are equally important. On the one hand, it depends on the orbit of the comet and, on the other hand, on the properties of the nucleus.

Concerning the orbit, it is essentially a question of how close to the Sun the comet gets. The closer the comet is to the Sun, the larger the chance for the comet to reach high brightness. This behaviour has a simple explanation: the comet nucleus contains volatile compounds. When the comet approaches the Sun, these compounds evaporate in the solar heat. It is this evaporation that fills the space around the nucleus with brilliant gas and dust.

The role played by the properties of the nucleus is considerably more complicated. The size is, of course, important, because most comet nuclei have too small a surface to exhume the enormous amounts of gas and dust which characterise the brightest comets. But small comets may too sometimes flare up and become hundreds of times brighter than usual. There seem to be multiple reasons for this, and these are connected to another important property: the presence of volatile substances.

In frozen form, volatile substances appear to be ice whose main constituent is water, i.e. H_2O molecules. But this ice is intimately mixed with other, much less volatile material. This involves carbon rich, organic compounds and different rocky minerals. As time passes and the comet goes round and round in its orbit, the heat from the Sun has an influence on the surface layer of the nucleus. As the surface temperature increases, the ice is vaporised and the vapour leaks out. One may thus expect the surface layer to become enriched in rocks and carbonaceous compounds, and all studies of comets support this. Hence, the freshness of the surface is another important factor. But the size and the freshness alone are not enough to understand all the variations in the activity and brightness of comets. This issue is not fully investigated and remains a subject of continued research.

1.2. The Nucleus — From Hypothesis to Proof

Our basic ideas about how the nucleus behaves manage to explain the celestial phenomenon that we call a comet. But this theoretical model of the nucleus only explains what we see. How do we know that comet nuclei exist in reality? Let us start with a historical summary.

In the 17th century, Isaac Newton's work on gravity and celestial motion established the foundations of calculating the orbits of heavenly bodies around the Sun. He was also among the first to utilise this knowledge in practical application to comets — e.g. the Great comet of 1680. But in these models, the celestial body is presumed to be point-like. In practice, the calculated orbit applies only to the centre of mass of the comet but not necessarily to its peripheral parts, such as the case may be for an enormously big comet. Hence, it is natural that Newton's own picture of the comet reminded of that of our days, where the mass is confined to a small nucleus.

Since then, Isaac Newton's version of the comet has been common knowledge. But the scientific method is about seeking out the weaknesses of theories rather than exaggerating their strengths, so that the theories can be replaced by something better. For want of proof of the existence of the comet nucleus, this theory has faced many sceptics. However, the absence of counterproofs was also striking. One exception occurred in the 1860s. The Italian astronomer Giovanni Schiaparelli showed that two well-known, recurrent meteor showers — the Perseids and Leonids — are caused by the Earth crossing streams of small particles in orbits similar to those of known comets. This made many — though far from all — imagine comets to be simple condensations without any real nuclei. This idea was also supported by the mysterious fact that some comets which had once been seen as permanent fixtures of the sky, suddenly disappeared. Maybe this was because the condensations had simply dissolved?

As a counterargument, the champions of the nucleus could point to cases where comets had split into several parts. This was the case of the lost comet Biela, a faithful re-visitor with a 6.6 year orbital period from its discovery in 1772 until its return in 1846, when suddenly two comets were observed close to each other. These were observed again during the next return in 1852 but were never seen again thereafter. The only reasonable explanation seems to be that the comet split up into increasingly smaller and smaller pieces until the Earth was washed by a magnificent rain of meteors in 1872 when it again crossed the comet's orbit. A comet's nucleus may possibly split up in this way, but it is hard to imagine a simple condensation in a cloud of particles doing so.

That the idea of a comet nucleus was far from dead is shown clearly by Jules Verne's novel written in 1877, *Off on a Comet*. In this book, we follow the French army captain Hector Servadac, who is torn off the Algerian coast by a passing comet and travels with the comet for one orbit around the Sun until two years later the comet returns to the Earth and Servadac manages to get back to Algeria. This story is, of course, a wild fantasy, but it shows that Verne must have had a big, solid nucleus in mind.

Much later, at the time of my own birth, the first physical theory of comet nuclei was formulated by the American astronomer Fred Whipple (see Fig. 1.2). According to Whipple, the nucleus is just the kind of intimate mixture of ice and dust I described earlier. The size would, in typical cases, be estimated at one kilometre, and it would later be described in popular terms as a dirty snowball. With this model, among other things, Whipple was able to explain some curious behaviours characteristic of comets. Their motions around the Sun are perturbed not only by the gravity of the planets but also by another force, whose nature had been debated for more than a

Fig. 1.2. Fred L. Whipple (1906–2004), American astronomer, considered by many as the father of modern cometary physics (to the left), and Horst Uwe Keller, leader of the ESA/Giotto camera team and main responsible for the images that served to verify Whipple's idea about the comet nucleus. Courtesy H. U. Keller.

hundred years. In Whipple's version, the disturbances are due to a jet force caused by the water vapour flowing out from the nucleus in an asymmetric manner. The gas stream is roughly directed toward the Sun, originating from the warm, sunlit side. The effect is like what you see if you inflate a toy balloon and let it go without tightening. The air streams out of the hole, and the balloon shoots off in the opposite direction.

When I studied astronomy in the 1970s, we were taught that Whipple's work from 1950 was considered as the foremost authority, but still there was some doubt. The sceptics did not have much to argue except the attitude that "I shall see it before I believe it", but as yet nobody had *seen* a comet nucleus. Any observations were hard to interpret, and while some claimed to have observed pure, bare comet nuclei, others could show that such claims were overly ambitious.

The issue could not be settled until humanity was able to send spacecraft into the immediate surroundings of a comet nucleus. This

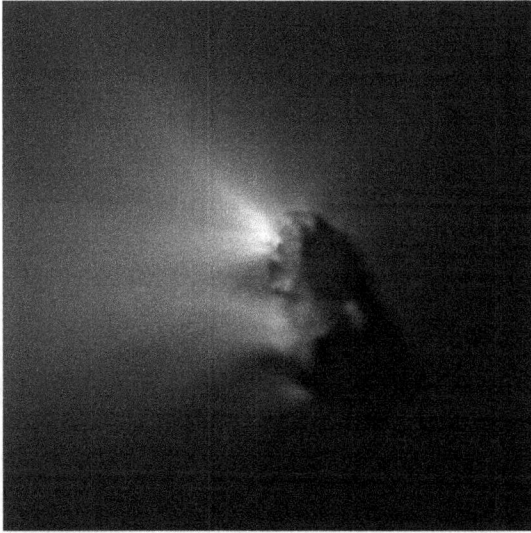

Fig. 1.3. Composite image of the nucleus of Halley's comet, acquired by the Halley Multicolour Camera on board the ESA Giotto spacecraft in March 1986. Reproduced with permission from ESA and H. U. Keller. © ESA/MPAe.

happened with the greatly awaited return of Halley's comet in 1986. In March, three probes flew closely past its nucleus to image it. The best pictures came from the Giotto probe, which was constructed by the European Space Agency (ESA). Here, the nucleus appears as an elongated, potato-like celestial body with dimensions about 15×8 km. A well-collimated outflow of gas and dust is seen on the sunlit side of the nucleus (see Fig. 1.3).

It is fascinating to imagine that we are here looking at the source of the comet that adorned the sky in 1986 — for once not particularly impressive — which has flown by on regular recurring occasions throughout human history. The previous visit was in 1910 when it came much closer to the Earth and therefore shone much brighter. The renaissance artist Giotto di Bondone is believed to have had it as a model when he painted the adoration of the magi in the Scrovegni Chapel in Padua at the start of the 14th century. According to the Bayeux tapestry, the same comet troubled King Harold as it was seen shortly before the Battle of Hastings in 1066.

These images of the Halley nucleus from 1986 are nothing compared to those provided by later space missions. Furthermore, the existence of the nucleus was hardly surprising. Yet, they have reached almost iconic status in comet research. These images settled the clinching question about the nature of comets and contributed strongly to our dawning understanding of the properties of comet nuclei.

1.3. Large-Scale Properties

The first question concerning the structure of a comet's nucleus is whether the nuclei are simple (i.e. monolithic) or consist of two or more components. The stars of the Milky Way are often in binary or multiple star systems, and the planets in the solar system mostly have their own satellites. So, what about the comet nuclei? So far, no nucleus visited by spacecraft has presented any large satellites. But this does not tell us much. The gravity of the explored nuclei is so incredibly weak that the previously mentioned jet force should make any components drift apart in a very short time. This is verified by split comets — no surviving double nuclei have been observed. On the other hand, as we shall see in what follows, the previously observed monolithic nuclei often appear to be contact binaries: it is possible that they once arose from mergers of several smaller parts.

The second question concerns the mean density of the nuclei. Density is obtained if one divides the mass by the volume. However, in a comet this is easier said than done. If one has a series of good images of a nucleus, one can crudely estimate the volume, and if this is unavailable, one can sometimes make an even cruder estimation. But the situation with the mass is different — how does one weigh a comet nucleus?

Using the gravity of the nucleus is nearly unfeasible since it is too weak to be measured. All space probes but one passed their target comets at such high speeds that no gravitational influence on their motion was measurable. Instead, we rely on the jet force.

The main observable effect of this force is that the comet does not arrive on time after one revolution about the Sun. By "on time" we mean the exact time calculated with regard to all the gravitational effects of the planets and the Sun. The discrepancy is usually called "non-gravitational effect", and its cause is the jet force. Using Newton's Second Law, which states that the force equals the mass times the acceleration, the measured delay is an expression of the acceleration imparted to the nucleus. One only has to find a corresponding expression of the force in order to calculate the mass.

Halley's Comet arrived about 4 days too late after each orbit. When the comet rounded the Sun in 1986 and its passage was nearing completion, it was more or less clear how much water vapour had flown out during each orbit. I then set out to theoretically compute the average velocity of these molecules as they left the nucleus. This gave me the necessary estimate of the jet force. My results were quite surprising. My estimate of the mass together with the volume of the nucleus estimated from the close-up images showed that the nucleus had a very low density. Within the frame of the uncertainties, this could be as low as one-tenth of the density of compacted ice.

I was the first to present and publish this internationally. But I do not think I really understood what I ventured into. The older among the world's leading authorities took a positive stand and considered that I had discovered something important. I especially remember Fred Whipple's commendatory words and that I gained a personal friend in his somewhat younger colleague Mayo Greenberg. On the other hand, I had a lot of trouble defending myself against attacks from younger scientists on both sides of the Atlantic, who tried to show that I was wrong or had greatly underestimated the uncertainties. Other papers were published with results at odds with mine, and many regarded the low density of comet nuclei with scepticism.

In retrospect, however, I can establish that the more recent results on cometary densities tend to agree with my own. And whatever the case might be, I opened a new door concerning opinions on the structure of comet nuclei. Since then, "a high porosity with gas flowing in and out through the pores" has become an attribute of most other scientists' theories.

If we imagine comet nuclei as drifts of newly fallen, dirty snow instead of hard-squeezed snowballs, the question arises as to how strongly they are held together. This can be studied by observations of split comets. Such observations have been available for almost 200 years, and as time passes these have grown quite numerous. It turns out that the splits are of two kinds: provoked and unprovoked. In the first case, we have to deal with an external influence and in the second case we don't know what it is.

By external influence we mean the so-called tidal force that the comet is subject to when it passes too close to the Sun or a planet. One may compare this to the tides that the Moon causes on the Earth's oceans. By the action of lunar gravity, the Earth is stretched out on both sides: the one facing the Moon and the one turned away. The bedrock is influenced as well, but the effect is very small (there is no risk for our planet to be torn apart). But the tensile strength of comet nuclei is comparatively so small that they may easily be split in the vicinity of large masses — the Sun or Jupiter in the cases we know of.

However, most splits occur without any obvious external reason. There are theories claiming that the gas flow leaving the nuclei makes them spin faster and faster until the centrifugal force surpasses the tensile strength, but there is no proof. In any case, it is clear that the fragments drift apart very gently as if they were barely stuck together from the beginning.

All in all, there is but one comet for which the split could be interpreted in terms of material strength in a quantitative way. This concerns one of the most famous comets of all time — Shoemaker–Levy 9. I feel personally involved in this case, too — at least marginally. By the end of the 1980s, together with my PhD students Mats Lindgren and Gonzalo Tancredi, I investigated the motion of a small comet called Helin–Roman–Crockett and found that this is sometimes captured as a distant satellite by the gravity of Jupiter.

We discussed this in our research group at Uppsala Observatory and concluded that such captures may be very common and that it could pay off to search the photographs of Jupiter's surroundings for the comets in question. Together with my colleague Claes–Ingvar Lagerkvist, Mats and Gonzalo acquired observing time at the European Southern Observatory (ESO) in Chile. Mats went on the trip as an observer and took pictures with a so-called Schmidt telescope. In March 1993, he discovered something very strange: something that might be reminiscent of a comet but extended into a short stroke. Could this be a comet that moved so incredibly fast that its image turned into a stroke? No, this was impossible, so what was it — something real or just a ghost image caused by the telescope optics?

Before Mats had a chance to explore the issue, the same phenomenon was seen at the Mt Palomar Observatory in the USA by Eugene and Carolyn Shoemaker (together with David Levy) in connection with a project that had nothing to do with comets. They were sure that this was a comet and reported it immediately. Since this was their ninth nameable comet discovery, the comet got the name Shoemaker–Levy 9. But imagine if they had been clouded out and Mats could have unravelled the issue in peace and quiet? If so, this famous comet might have carried the name Lindgren.

Shoemaker–Levy 9 was exactly the type of comet that Mats was looking for: a temporary Jovian satellite. But during its preceding orbit around Jupiter, it had penetrated so close to the planetary surface that it was split by tidal forces. This happened in July 1992. From one single comet, there appeared more than a dozen, which were marching on toward the next encounter with the gas giant (see Fig. 1.4).

It was this next appointment in July 1994 that made the comet famous. All the fragments — one after the other — collided with Jupiter. The event was predicted, and both scientists and laymen alike were sitting watching in excitement as the spectacle occurred. Yet, I claim that the most important knowledge acquired by humanity from Shoemaker–Levy 9 did not come from those collisions but from the preceding split that nobody had observed.

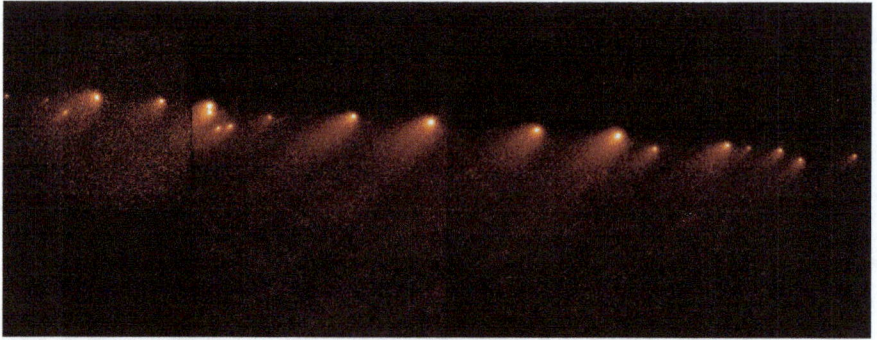

Fig. 1.4. A mosaic of images from the Hubble Space Telescope, showing the whole train of fragments constituting comet Shoemaker–Levy 9 on May 17, 1994, two months before these plunged into Jupiter's atmosphere. Credit: H. A. Weaver and T. E. Smith (STSci), NASA.

Studies of this split using physical models and computer simulations, performed by Willy Benz and Erik Asphaug, showed that the original comet nucleus was likely quite small, probably with a low density, and that its material strength was nearly nil — it was held together primarily by its own, minimal gravity!

1.4. The Chemistry of Comets

The scientists' notion of the detailed chemical composition of comets has undergone large changes. An early indicator came from exploration of Halley's comet, whose nucleus was much "dirtier" than one had believed. One already had a good idea of how much gas it produces thanks to earlier telescopic observations. But instead of a relatively small nucleus made of almost pure snow, the comet proved to have a much larger nucleus, where on average each square metre of the surface only produces one-tenth of the expected amount of gas.

It was natural to infer that the surface had been dirtied by non-volatile residues from earlier passages near the Sun. If this was the case, almost the whole nucleus — i.e. everything below a thin surface layer — might consist of almost pure snow. However, the *in-situ* measurements by the space probes showed that the so-called *dust* of the comet — i.e. the solid particles dragged along in the gas flow from the nucleus — mostly consist of massive clumps, which are far too large to have been seen in the earlier observations. As a consequence, the total flow turned out to contain at least as much dust as gas. If this behaviour is typical and not just a temporary anomaly, one may naturally conclude that the dust is a dominant constituent of the nucleus. Thus, as the comets would not be essentially icy, it was suggested as a joke that the accepted description of comet nuclei as *dirty snowballs* should be changed into *snowy dirtballs*.

From this new perspective, the comet dust increases in importance. To understand the nature of comets, one should concentrate one's efforts on the dirt rather than the snow. Indeed, studies of the dirt have led to interesting results. The dirt has proved to consist of two very different types of material, which are intimately mixed and occur in similar proportions. One type is rock and is built of the same silicate minerals dominant in Earth's mantle. The other can be described as complex, carbonaceous compounds adhering to organic chemistry. The analysed grains in Halley's comet of the latter nature were named CHON particles since they essentially consist of the chemical elements carbon (C), hydrogen (H), oxygen (O) and nitrogen (N). It has not been possible to certify which molecules contribute precisely, but I remember that one of the scientists, Jochen Kissel, with his characteristic humour said that the environment that the Giotto probe crossed rather reminded him of the exhausts from a diesel engine.

Globally, then, the comet is a porous mixture of rocks, CHON material and ice. The abundance of the ice is difficult to establish, but at least it does not seem to dominate. There is a theory elaborated by the American astrophysicist Mayo Greenberg (see Fig. 1.5), which fits well with several observations of comets. In brief, this involves the following. Microscopic dust grains are formed in the atmospheres of stars and are blown out into the vast space between the stars in the Galaxy — the so-called interstellar space. Out there, they sometimes pass through extremely cold and dense environments, and simple molecules arise built of the most abundant reactive elements, i.e. the previously mentioned CHON elements. These molecules freeze into

Fig. 1.5. J. Mayo Greenberg (1922–2001), world-leading expert on the physics and chemistry of interstellar grains. The picture shows Greenberg holding a model that illustrates his concept for the composition of comet nuclei. Courtesy Leiden Observatory, Laboratory for Astrophysics.

ice, forming a mantle around each rocky grain. As the molecular cloud is dispersed, it is heated up by starlight, and the icy mantles reach temperatures high enough for the molecules to react with each other. Consequently, large quantities of chemical energy are liberated, heating the grains further. The reactions then proceed explosively, leaving behind a mantle of complex organic substances. These grains continue their journey through interstellar space until, eventually, they end up in a molecular cloud where stars with planetary systems are born — e.g. the solar system.

From this perspective, one may possibly imagine that the comets were built up in the infancy of the solar system by interstellar grains, where the rocky core and the organic mantle were swept into an icy mantle from the cloud where the Sun was born. This idea had many proponents during the 1980s and 1990s, but it was neither confirmed nor rejected and may thus far be regarded as an unconfirmed hypothesis. It is however interesting — not least because the comets would then be largely built of material older than the solar system. If we can reach them, we may be capable of studying the very material that the solar system was built of.

What about the ice of the comets? As yet, we cannot make statements as specific as for the dust. The compounds in the ice are so volatile that they can easily both evaporate and re-condense if the temperature varies. Whether this happened or not when the solar system was formed, is both an important and unanswered question. But whatever the answer is, a great deal of interest has been focused on which specific compounds are there, in which form and in which proportions. Gratifyingly enough, this puzzle has been solved considerably during the last 30 years, firstly thanks to new observational techniques, and secondly, because Nature has invited us to a few very bright and easily observed comets. I'm thinking in particular of the two stunners in the 1990s: Hyakutake and Hale–Bopp.

There is no doubt that the water molecule (H_2O) is the most common. But this does not tell us much, because anything else would be a great surprise. Water is the most abundant molecule everywhere in the universe, where the environmental conditions allow it to exist. In addition, H_2O evaporates at a higher temperature than ices of other volatile substances. Thus, there are many situations where the temperature is such that the water is frozen but not the other compounds. Hence, these other compounds can act as temperature indicators. Which ones do and don't exist in the ice of comets tells us something about where and when the comets must have been formed, since the correct temperatures must have prevailed at the correct time and place.

The two oxides of carbon (carbon dioxide (CO_2) and carbon monoxide (CO)) are of particular interest. The behaviour of carbon dioxide is familiar to anyone who has seen dry ice boil. It stays frozen only below $-78°C$. Many also know that liquid nitrogen is extremely cold — not far from absolute zero — and the case would be about the same with carbon monoxide. Observations have shown that both carbon oxides are among the most abundant species after water. But carbon monoxide seems to vary considerably more from comet to comet, as if the places where the comets were born were so cold that the carbon dioxide always froze but the carbon monoxide only sometimes froze. This has been suggested to be the case, but I think that some scepticism is warranted.

One ground for scepticism is that we do not know enough about the form in which the two kinds of molecules are stored inside the comet nuclei. Either the ice consists of different components stored at different temperatures — i.e. H_2O ice, CO_2 ice and CO ice — or it is exclusively a question of H_2O ice where the other species are trapped as guest molecules. The second alternative mainly concerns the carbon monoxide, but we do not know how often it occurs. In any case, it is clear that the molecules leaking out of a comet nucleus on one given occasion are not necessarily representative of the ice as a whole. We find an obvious example with comet Hale–Bopp (see Fig. 1.6).

Hale–Bopp was discovered during the summer of 1995, almost two years before it came closest to the Sun. At that time it was positioned more than seven times further from the Sun than the Earth's orbit, where it was far too cold for the H_2O ice to sublimate. But the comet was still active, and the activity was driven by an intense flow of gas from the nucleus. It turned out that CO was a strong contributor to this gas. We now know that CO dominated over H_2O until Hale–Bopp reached four times the Earth's distance from the Sun. On the way out, CO reclaimed dominance as the distance reached three times that of the Earth.

It may have been the case that the flow of CO was caused by the H_2O ice crystallising at very low temperature (about $-140°C$) without evaporating but allowing trapped CO molecules to escape. Or perhaps, it was instead CO ice that sublimated at an even lower temperature. I would rather bet on the first option, but the issue is not entirely explored. In any case, the enormous flux of H_2O as the comet rounded the Sun showed that the water ice is predominant in this comet.

Fig. 1.6. Photo of comet Hale–Bopp taken on March 29, 1997. Credit: Philipp Salzgeber. License: Creative Commons Attribution-Share Alike 2.0 Austria.

The list of other molecules observed in comets has grown rather long with time. One finds, for instance, both oxygen and nitrogen gas but also methanol, acetylene, formaldehyde, hydrogen cyanide, ammonia, methane, ethane and hydrogen sulfide. Not any nice gas mix, one may think, but this is the reality of comets of our solar system and generally in the regions of the Milky Way where stars are born. This once more indicates that comets contain samples of the solar system's original material. The gases may have been released from evaporating ice and re-condensed, but they were not transformed by chemical reactions.

1.5. Wild 2, Tempel 1 and Hartley 2

New methods to observe comets include larger and more sensitive radio telescopes for high frequencies, infrared telescopes on high altitude aircraft, and all kinds of satellite telescopes. But a special place of honour still belongs to the space probes. After the instant fly-bys of Halley's Comet, another comet named Borrelly was visited by the NASA probe Deep Space 1. I will now give brief accounts of all the following projects, ending with the champion of them all, ESA's Rosetta probe.

The targets of all these probes were comets with orbits in the innermost part of the solar system and short orbital periods. The first was the NASA probe Stardust, which reached its target, comet Wild 2, in early 2004 (see Fig. 1.7). The nucleus of Wild 2 is less elongated than the Halley nucleus, and on its surface we see large, round basins. These may possibly be traces of collisions from when the comet was moving in a region further from the Sun. In fact, we know that it had such orbits recently — until 1974 — so the scars

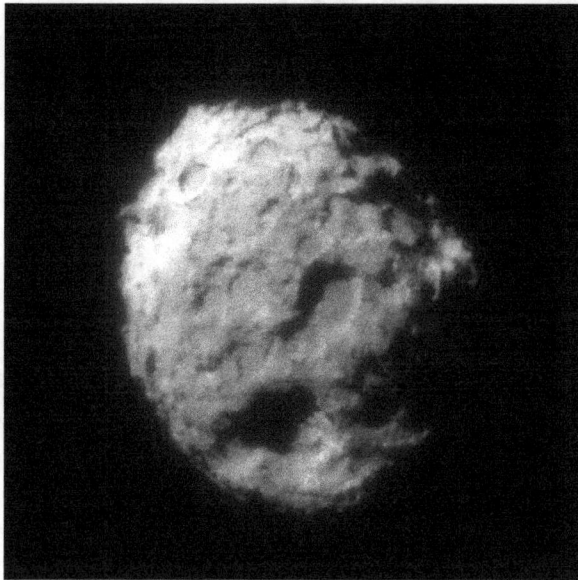

Fig. 1.7. The nucleus of comet Wild 2, imaged by the Stardust camera. Credit: NASA/JPL-Caltech/University of Washington.

from its past have not had the time to be eroded away. I shall return to what happened in 1974 in Section 4.3.

But Stardust's great achievement was something else. It carried a kind of dust collector made of extremely porous silica gel, which was exposed to dust particles as the probe raced through the comet head at a high speed. The harvest was many thousand grains of comet dust which travelled onward with Stardust back to Earth, where, in January 2006 after a perfect landing in the Utah desert, it was brought to the laboratories for analysis while the probe rushed on toward further adventures. To me, this is one of the most impressive feats in the history of space research. This being said, it is a fact that the returned samples suffered from natural limitations. The grains had not exactly made soft landings in the silica gel: they hit with a speed of 6 km/s. Much of the material was thus vaporised, and what was left was dominated by the most resistant, rocky minerals. This selection effect presents a serious drag upon the interpretation of the results. I therefore leave these aside, although they have been much discussed by scientists and are considered to be of great importance.

The next probe also came from NASA. This was named Deep Impact after the well-known 1998 disaster movie about a fictional comet impact next to the US. But this time it was humanity that hit a comet — not the other way round. The victim of this experiment was comet Tempel 1, which was known about since the mid-19th century — however, it was not badly injured. The aim was to study the comet nucleus at depth instead of just the surface. It was known that the surface of the nucleus may differ greatly from the interior after having grilled and sweated in the solar heat several times. The Deep Impact probe fired a massive projectile on a collision course with the nucleus as the probe passed nearby. This impacted the comet as foreseen, causing a small explosion, whereupon scientists studied the ejected material.

This happened on July 4, 2005, and I don't think this date was chosen arbitrarily. In any case, it was a nice spectacle, which was carefully observed both from Deep Impact itself and from telescopes on the Earth, on satellites, and in the 80 million km distant Rosetta probe. Among the noteworthy results, the excavated material was fine-grained, rich in water and carbon dioxide, and remarkably rich in dust. It was not possible to directly determine the size of the crater caused by the explosion, but the answer came six years later from another space probe.

This was the retired Stardust which, after the delivery of the captured material from Wild 2, had enough remaining fuel to be steered to Tempel 1 for a final task. Fortunately, the camera was still in good shape, and the pictures taken of the Tempel 1 nucleus were quite important. Among others, the diameter of the crater caused by Deep Impact was estimated at 150 meters. The combination with the old pictures from 2005 also provided a stereo effect that made it possible to determine the size and shape of the nucleus accurately.

In addition, Deep Impact left behind at least two further, important results. The surface temperature of the nucleus was mapped and found to be so high that no cooling effect from the sublimation of the ice could be noticed. The conclusion was that the gas flow in Tempel 1 emanates from a non-negligible depth below the surface. Furthermore, it was found that the icy spots seen on the surface were so small that they could contribute only a very small fraction of the observed water vapour. Finally, the images from both 2005 and 2011 showed that the nucleus seems to be constructed of interesting, sequential layers. Sometimes these layers are reminiscent of bathing hoods covering large parts of the nucleus. One popular theory, launched by Mike Belton, claims that these arose when the young comet nucleus was bombarded by smaller bodies, which impacted at a modest speed, were pulverised and flew away in all directions from the point of impact. The model in question received the name "talps", i.e. "splat" read backwards.

The Deep Impact probe too was directed to another comet. This was the smallest among all nuclei so far visited by spacecraft, associated with comet Hartley 2. Malcolm Hartley had discovered it in Australia in 1986 using a Schmidt telescope belonging to the Uppsala Observatory. Concerning the nucleus, we can say that it is somewhat of a joker among the space probe targets because of a few unexpected and surprising properties (see Fig. 1.8).

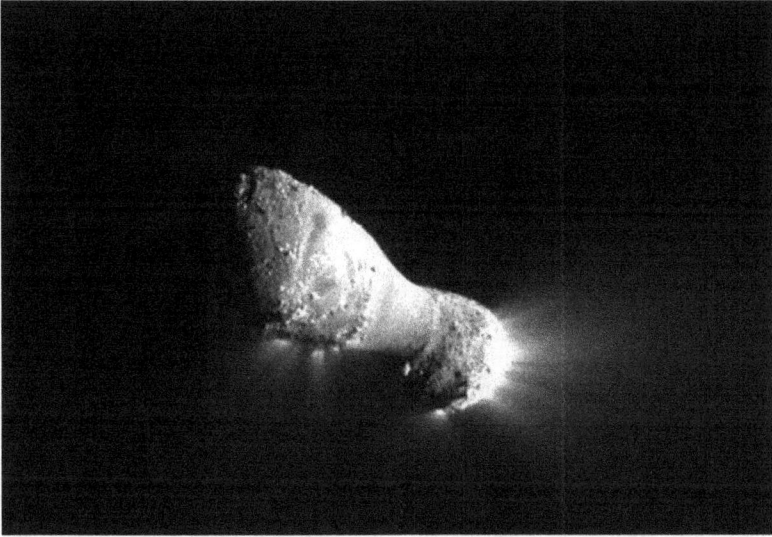

Fig. 1.8. The nucleus of comet Hartley 2, imaged by the Deep Impact spacecraft within the EPOXI mission. Credit: NASA/JPL-Caltech/UMD.

On the images, one can clearly see that the nucleus has two parts or lobes that are joined by a "waist" in a way somewhat reminiscent of an ant's body. Even though it was possible to speculate about such a structure in other comets including Halley, this was the first time it appeared so convincingly.

The next thing to consider is how differently the two lobes behave. The pictures were taken as the comet was at its closest to the Sun, and the solar illumination was very strong. But while the smaller lobe is spurting out gas and dust, the waist and the bigger lobe seem tranquil and indifferent. The explanation appears to be that the surface layer of the smaller lobe contains a very large amount of carbon dioxide. When this evaporates and flows out into space, it drags along the rest of the material in the form of grains that we call "dust". But this dust is not like the dust of usual comets — it mostly consists of H_2O ice.

The H_2O ice evaporates as well of course, because the solar heat is largely sufficient. The amount of water vapour produced is enormous and exceeds what one would have if the whole nucleus were covered by pure H_2O ice. This circumstance had already been observationally established, and comet Hartley 2 had thus been called hyperactive. Concerning Halley's comet, I mentioned earlier that its production of water vapour was only about 10% of what a pure snowball would yield. For Tempel 1 it was even less. But for Hartley 2, the value is more than 100%, which is very uncommon. One can now guess that the reason is the same also for other comets with similar values.

Another interesting observation from Hartley 2 is what may be called ice dumping. What happens to those icy grains that are so big that the flow of carbon dioxide does not manage to drag them into space against the gravity of the comet? These naturally fall back to the nucleus and land somewhere else. The waist region is particularly attractive for these remnants since it has the lowest gravitational potential. True enough, this seems to be covered by dumped icy chunks originating from the small and active lobe. The waist is thus the source of part of the water vapour, as confirmed by the observations. The same type of ice dumping would also be observed during the latest exploration of a comet nucleus by a space probe, i.e. Rosetta.

1.6. Rosetta

Discussions of precursors to the Rosetta mission began during the 1980s. The NASA plans dealt with project CRAF, i.e. *Comet Rendezvous Asteroid Flyby*, while ESA imagined something even more ambitious called CNSR: *Comet Nucleus Sample Return*. In the end, none of these projects were realised, but ESA went ahead with the new project Rosetta, which in large part was an upgraded version of CRAF. In the early 1990s, it was clear that this would be one of ESA's cornerstone projects awaiting the millennium shift, and discussions started around the instrumentation. The basic idea was, with the aid of a comet nucleus, to learn to read the earliest history of the solar system roughly like the trilingual Rosetta stone allowed people to interpret hieroglyphs and read about the history of the Egyptian civilisation. Comet nuclei are our most important witnesses, being ancient and having been kept in cold storage far from the Sun ever since their formation.

The strategy of the Rosetta mission was to place the probe in a similar orbit around the Sun to that of the comet to be explored. In this way the probe could approach the comet far from the Sun at a very low speed, brake further and then follow the comet on its way toward the Sun and, hopefully, also part of the way out after rounding the Sun. Moreover, a robot would land on the surface of the nucleus and transmit its measured data to the mothercraft for further delivery to Earth. One would thus be able to study in some detail all that happens to the comet nucleus during this journey and study what leads to the typical comet that we see in telescopes from the Earth. Lo and behold, this would be done in peace and quiet in contrast to the earlier, almost instantaneous fly-bys.

But to place a space probe in the same orbit as a comet is not easy. It takes considerable time since one has to perform several close encounters with the Earth and Mars, where planetary gravity is used to deflect the probe into more and more comet-like orbits. In Rosetta's case, these manoeuvres took 10 years, and everything on board had to be constructed so that the instruments would stand the hardships of space — such as corpuscular radiation — for a long time.

There were not many comets to choose between which had a suitable orbit for this particular mission. The first choice was comet Wirtanen, which became increasingly worrying as the date of launch approached in the winter of 2003. Its nucleus seemed uncomfortably small, hyperactive and unsteady. But, still, everything was ready and prepared, until a launch of the ESA rocket Ariane 5 failed in the autumn of 2002. This would have been the last such launch before the launch of Rosetta and, awaiting the investigation of the failure, it was decided not to proceed. Hence, the trip to Wirtanen was called off, and a new target comet had to be found for launch on a later occasion.

Fortunately, there was a perfect alternative, namely, comet Churyumov–Gerasimenko, and for that the launch would only be delayed by one year. Thereafter, things went perfectly on track with very few mishaps. The worst of these was the unsuccessful touch down of the lander Philae, hampering the use of this robot. All in all, Rosetta remains one of the most successful space projects of all times. It is very hard to give a correct and impartial description of all the results, and I won't even try. I'll be content with illuminating some things that I find relatively secure and interesting, but I cannot claim to speak for the world's collected expertise.

My own impression of participating in the Rosetta observations — in my case within the camera team — can be described with a metaphor. Imagine that you were taken on a tour of Paradise! You would have your expectations and prejudices, but you would be totally unable to directly grasp what you'd see, since it was so far from your experience. After the tour you might discuss with your companions what Paradise really is, but you would not get anywhere. Every person would have their own idea of Paradise and it would be hard to share the same perspective. Of course, the Rosetta experience was not so bad and nightmarish as this, but yet, we who participated got to see things that no one had seen, and it was not easy to agree on what they meant.

Starting with the shape of the nucleus. Something as weird as what we saw was beyond my imagination (see Fig. 1.9). The nucleus is clearly bilobate. It has two parts, which seem to barely stick together by a narrow joint. To describe this appearance, everyday comparisons have been used, and the most common is a rubber duck. The smaller part is the head, the larger part is the body, and the joint is the neck. The nucleus rotates with a period between 12 and 13 hours — this period changes from orbit to orbit due to a torque that arises, as the gas streams out from the sunny side of the nucleus. The spin axis runs across the nucleus through the body in the vicinity of the neck.

There are in principle two ways to explain the origin of this shape, and I call these the apple snufkin theory and the snowman theory. According to the first, the neck has been hollowed by erosion through ice evaporation out of an originally smooth and elongated nucleus. One often reads about this possibility, but on closer inspection, it does not work. We instead have to concentrate on the second, where

Fig. 1.9. The nucleus of comet Churyumov–Gerasimenko, imaged by the Rosetta Navigation Camera on September 19, 2014. Credit: ESA/Rosetta/NAVCAM, CC BY-SA IGO 3.0.

the two parts have been joined together gently like the head and body of a snowman. How this occurred and in which connection will be dealt with in Chapter 6.

One may also speculate about the future fate of the comet nucleus. If the joining happened very gently, the joint is not very strong. Whether it survives the shear that arises as the two parts are turned in different ways by the outflowing gas is another issue. There may possibly come a day when comet Churyumov–Gerasimenko is but a memory and two new comets appear. The same event may have happened to another comet in 1845 (see Section 4.5). The two comets Neujmin 3 and Van Biesbroeck have different orbits, but in 1984 it was discovered that both of these were strongly perturbed at close encounters with Jupiter in 1850. Before these encounters, they were very much alike. In March 1845, the positions of the two comets practically coincided, and it seems likely that at that time, these originated as fragments of a split parent.

By combining close-up images of the nucleus taken from different directions, we have constructed a detailed, three-dimensional model, which yields the volume with good accuracy. The mass could be measured as well using the radio signals of the Rosetta probe and providing insight into how the gravity of the nucleus influenced the motion of the probe. The resulting mean density is as low as $0.53\,\mathrm{g/cm^3}$, which means that the nucleus must be highly porous. The true value of the porosity depends on assumptions about the mixture of materials constituting the nucleus. This involves some uncertainty, but a porosity exceeding 70% is a safe bet. Hence, most of the volume in the nucleus is occupied by vacuum and only a lesser part by material. There are also good signs that the nucleus is fairly homogeneous. Large hollows do not seem to exist, but the porosity is present everywhere on a small scale to a similar extent.

The equator of the nucleus is of course defined by a plane perpendicular to the spin axis. In this comet, the equatorial plane is inclined by as much as 52 degrees to the plane, where the orbit is situated. This means an extreme amount of seasonal variation, and midnight Sun and noon darkness are common. In addition, the orbit is quite elongated so that the distance to the Sun varies strongly during the course of the orbit. This is quite common among comets at large and thus is not specific to the Rosetta comet. Something that may be specific to the Rosetta comet is that the northern hemisphere of the nucleus has a very long and cold summer with the Sun shining for a long time — though weakly — while the southern hemisphere has a long polar night that is compensated by a short but very intense summer when the Sun shines high in the sky.

As a consequence, surface material is transported from south to north during each southern summer. The southern ground absorbs lots of solar energy. This breaks it apart, and the fragments sail away in the gas flow coming from underneath. However, they cannot escape the gravity of the nucleus and fall back in all kinds of places. On the northern side it is cold and quiet, and the material is gathered there, waiting for the next northern summer. Rosetta observed the effects of this transfer in different ways.

Mapping the geology of the nucleus and registering the events taking place in detail has its interest of course and may be both entertaining and impressive, but this was not the reason that Rosetta was sent on its mission. The question to be answered was: how do comets work? It is not enough to just register what happens, we have to understand why. If not, Rosetta has not taught us what we hoped for.

Yet, I do not think that the adventure ends in such a disappointing way. As I indicated, we have to free ourselves of our prejudices and open up our eyes to the possibility that many things we took for granted were wrong. Here, I see two old dogmas that may have to be torn down. One is that ice is a dominant constituent of comet nuclei, and the other is that the prime driver of cometary activity is the sublimation of H_2O ice. Neither of these may be completely wrong, but they have to be taken with a large grain of salt. It is strange, but I may have to accept that two of the things that I once felt most sure about turned out to be quite unreliable.

A new picture is appearing. It is too early to judge if it tells us the truth, but it is definitely worth considering. An essential key is contained in the studies of the cometary dust grains performed by Rosetta from its vantage point in orbit around the nucleus. The low speed of the probe, the systematic monitoring of different aspects of the nucleus, and not least the long duration of this research, have opened new eyes to cometary dust grains. Thus, one could see particles both much bigger than ever before and a new kind of tiny agglomerates.

The largest blocks have volumes up to about one litre and, being rather compact, they have considerable masses. Beneath a dry crust they consist partly of ice. But the tiny agglomerates are completely dry and so porous that they cannot even be described as three-dimensional objects — they are actually fractals. We shall see in Chapter 6 that these fractals may represent one of the first stages in the origin of comet nuclei.

By comparing the amount of ice to the amount of non-volatile substances in the ejected material over time, as these emerge from the Rosetta measurements, it has been found that the latter type contributes a large majority of the comet material. This is a surprise to most of us. The ice that from the outset was thought to be dominant is now dispatched to a place in the dark. The contrast between the comets and some water rich asteroids is thus dimmed.

Another achievement was made by a camera on board the Philae lander. This imaged the surrounding terrain with a pixel size of about 1 mm in the course of a few days after the landing. The place in question has been named Abydos and is situated on the head of the nucleus. On some of the images there appears a grainy surface, where the grains lie side by side and are smaller than 1 cm. It is tempting to consider the whole nucleus to be built of such grains, and we shall see the reason for this in Chapter 6.

A new theory of cometary activity has been proposed by a research team including Jürgen Blum, Bastian Gundlach and Marco Fulle. According to this theory, the normal behaviour of comets can be explained by the mentioned grains functioning a bit like pressure cookers. The water vapour is trapped inside the grains so that the ice sublimates at a higher temperature than normally. The content of H_2O in the grains is low, but when the Sun shines brightly, the pressure still reaches high above the material strength of the grains. As a consequence, a surface layer of a few cm is crushed so that solid grains of all sizes up to this thickness are dispersed into space, and the water vapour streams out, helping them on the way.

This may explain why active comets such as Tempel 1 and also the Rosetta comet seem to have very little ice on the surface. The ice is hidden below a dry and very thin surface layer. But it is still the cause of the activity and can continue in this role as long as the grains are not completely dried up. But there is another part of the activity, which the H_2O ice cannot explain, namely, the large chunks that Rosetta observed. These must have been broken loose at a depth of about a decimetre, and at this level the pressure of water vapour is far too low. Instead, we must think of the most common, more volatile substances like CO_2 or CO. These too can build up a high pressure through the pressure cooker principle.

This happens mainly during the "hot" summer on the southern hemisphere. The ejection of big chunks was caught in action by Rosetta, as mentioned earlier. It is then an inevitable consequence that many chunks get dumped onto the northern side. These chunks

Fig. 1.10. The Hapi valley on the nucleus of comet Churyumov–Gerasimenko is seen on the left side and the Hathor wall rises from the valley on the right side. Image acquired by the Rosetta/OSIRIS narrow angle camera. Reprinted from Thomas, N. *et al.*, *Science* **347**, aaa0440 (2015) with permission from AAAS.

largely retain their original content of H_2O ice — only a thin crust is dried by ice sublimation. As the comet moves toward the outer parts of its orbit and the environment grows colder, water vapour flows out toward the surface of the chunks and once more freezes to ice (see Fig. 1.10).

This appears to have been the case for instance in the Hapi valley on the nucleus of the Rosetta comet — this was the first observed source of activity before and during the arrival of the probe in August 2014. This region is situated on the northern side of the neck and is surrounded by the kilometre tall Hathor cliff on the head and another sharp slope on the other side, where the body starts. Direct sunlight does not easily enter into this deep valley, but the surrounding scarps were both reflecting sunlight and being warmed by it, so that the bottom of the valley was in turn heated by their infrared radiation. Here, the ground was full of icy blocks that had been dumped as the comet previously rounded the Sun. As the temperature rose high enough, the pressure cooker started and the comet began to spurt out gas and dust.

In fact, there are great similarities between Churyumov–Gerasimenko and comet Hartley 2. In both cases, icy grains are detached by the pressure of volatile gases from one part of the nucleus to fall back onto another place, where the surface is quiet. The subsequent activity due to sublimation of the ice is certainly water driven, but it would not have existed without the volatile substances.

But the analogy with Hartley 2 has its limitations. On Hartley 2, the flying grains seemed to vaporise to such an extent that the comet became hyperactive, but on Churyumov–Gerasimenko the activity stays at a modest value of 5%. We do not know what discriminates the two comets, but the smaller lobe of Hartley 2 does not necessarily have a higher abundance of CO_2 even though this is the first idea that comes to mind.

Finally: what is the material covering the comet nuclei? The surfaces of all nuclei so far visited by spacecraft have turned out to be extremely dark, and this has generally been connected to the presence of carbon-rich substances akin to the CHON material discovered in Halley's comet. The nuclear surface of Churyumov–Gerasimenko was continuously monitored by an imaging spectrometer for near infrared radiation. This way, one was able to identify both several kinds of ice and organic compounds. The ice only peeped out locally and temporarily, but the organics were legion. Preliminarily, we are dealing with a relatively volatile mix. This might work as a source for the kind of CHON particles, which are split up in comets at moderate distances from the Sun and give rise to molecular fragments like CN, C_2 and C_3. These have been identified in comets for more than a century, but are usually considered to be of unclear origin.

1.7. Summary

After all these details, a summary of what we can claim comets to be is warranted. At depth, they are quite small heavenly bodies orbiting around the Sun and containing substances volatile enough to be called ice, which are able to vaporise in the solar heat in the innermost parts of the comets' orbits. The sizes of those visited by spacecraft are in the interval 1–10 km, but we are certain that there are both smaller and much larger objects among the other comets.

As far as we have seen, these bodies — the so-called comet nuclei — are highly porous and loosely consolidated. These have offered much evidence of their transiency and limited lifetime. Chemically, they consist of a conglomerate of substances with very different degrees of volatility. Among the components of the ice, we find everything from the dominating water to extremely volatile species. In the other, predominant part of the material we find everything from extremely refractory rocky minerals to organic, carbon-rich compounds nearly as volatile as water.

It seems as if the comet nuclei were once formed with all these components in an intimate, homogeneous mixture. This may represent a complete inventory of all that, in the newborn solar system, became the building material of planets and other solid objects. We also believe that the nuclei have been largely preserved over time — however, with one important exception, namely, the outermost surface layer. Here, the volatile material has largely escaped, so that the rest has become enriched as a kind of residue. An organic, carbon-rich slurry gives the nuclei a dark and neutral hue.

It remains a dream of many scientists to use the comets to fetch the pristine material that they were built from. But this appears ever more as a chimera. One would likely have to dig deep to access such virgin layers — in particular, if the most volatile species are of considerable interest, and unfortunately, this is probably the case. Yet, it is obvious that the Rosetta mission has advanced research in a decisive manner — yielding important clues as to how comets were once formed.

Chapter 2

Minor Planets and Dwarf Planets: Splinters and Survivors

It may be true that the nucleus is the only important thing about a comet, unless you value the comets for their beauty, but this does not negate the fact that the very definition of a comet requires the shining haze. If a celestial body in the solar system never exhibits anything of this kind, then, strictly speaking, it cannot be called a comet.

The term "asteroid" is sometimes somewhat carelessly used to refer to all such heavenly bodies, but it should rather be reserved for the members of the asteroid belt between Mars and Jupiter, and some others that originate there. We have no good word for the totality of all small bodies in the solar system that are not comets, regardless of where they are located. But the term "minor planet" is often used and may be accepted, as it does not imply any specific place of residence or chemical constitution.

Thus, minor planets can exist anywhere, and they can be built of anything. Nonetheless, there is a small group that has been isolated and given a special designation after a decision by the International Astronomical Union at its congress in Prague in 2006. The background of this decision was that the ninth planet, Pluto, was no longer to be called a planet and would hence automatically turn into a minor planet, unless something special was done. This was perceived by many astronomers — especially Americans — as a shameful demotion. The solution was to introduce a new category of bodies called dwarf planets, to which Pluto would belong. The somewhat curious meaning of the term dwarf planet implies that it must be bigger than usual minor planets but not big enough to, like the planets, emerge as the master on board in the zone where it moves. I refrain from discussing this any further.

Here, I don't see any reason to exclude or discriminate against dwarf planets, so I shall describe Pluto and other dwarf planets, for instance, the largest asteroid Ceres, as if they were simply the largest among the minor planets.

2.1. The Discovery of Asteroids

On New Year's night 1801, Giuseppe Piazzi stood observing stars at the Palermo Observatory. But one of the stars had an odd behaviour. It moved across the sky like planets do, but it appeared to be a star. It remained a brilliant point of light even at high magnification, while all the known planets could be seen as small disks. What would Piazzi call his object? He had no obvious idea but all the same reported his discovery.

The report caught great attention, in particular, among German astronomers, who had pointed out an interesting structure in the planetary system a few decades earlier. This was the so-called Titius–Bode law — a geometric series for the distances of the planets from the Sun. One place in this series was vacant, namely, the place between Mars and Jupiter, where the series predicted a planet that nobody had observed. Now the question arose, whether what Piazzi had observed was this planet. The first orbit determination seemed to confirm this, because Piazzi's object had exactly the predicted mean distance of 2.8 times Earth's distance.

This raised great enthusiasm, and had the Nobel Prize existed at that time, it would likely have gone to Piazzi. With the discoverer's right, he proposed to name the new planet Ceres after Sicily's guardian goddess. But clouds would soon appear at the horizon. At first, it was clear that Ceres was much smaller than the known planets. If it were to be called a planet, it would hence be a miniature planet — a real dwarf by comparison.

The next problem arose in 1802, when another object was discovered in another orbit with the same mean distance as Ceres. This was named Pallas. Within five more years, the third and fourth (Juno and Vesta) were discovered, too. Hence, instead of one real planet, it appeared that the gap in the Titius–Bode law was filled by a real swarm of miniature planets. This called for rethinking, and new ideas came forth. There is one thought that had a prominent place in the discussions and asserted itself for a long time. According to this, there had once been a large planet in the region in question, but it had exploded so that we now see only some of its debris. As we shall see later, we now know that the explanation has to be different.

What Piazzi and others did was initiate the discovery of the asteroid belt. After Vesta (1807), it would take until the 1840s before it continued, but since then the discovery process has been explosive. Ten asteroids were known prior to 1850. The number reached 100 in 1868, 1,000 in 1923, 10,000 in 1999 and 100,000 in 2005. At the time of writing, it is about 800,000. This means that, in the course of time, one has been able to discover ever smaller and fainter objects due to technical progress. To this comes yet another special circumstance, namely, the impact hazard to be described in Chapter 5.

2.2. The Asteroid Belt

The asteroid belt harbours millions of asteroids larger than 1 km in diameter. It stretches roughly from 2 to 3.5 times Earth's distance from the Sun (so-called *astronomical units*). The orbits are ellipses with moderate eccentricities, which are inclined by up to a few tens of degrees to the plane of planetary motion. This gives the region in question the shape of a somewhat flattened car tire.

The space embraced by the asteroid belt is immense. Hence, no immediate scrum prevails there, even though the asteroids are so numerous. Each of them is generally located very far from its closest neighbour. They are fairly evenly dispersed in space — there are no real holes or gaps in the asteroid belt.

On the other hand, there is something called a Kirkwood gap. The American professor Daniel Kirkwood discovered a strange phenomenon in the statistics of asteroid orbits as early as 1866, when less than a hundred asteroids were known. The mean distance from the Sun was rather evenly distributed over the previously mentioned interval, except for some values that were completely missing. He also noted that these values corresponded to simple resonances with Jupiter's orbital motion. With today's gigantic number of known asteroids, these Kirkwood gaps show up even more clearly and in more detail. It is a question of an integer ratio between the orbital periods of Jupiter and the asteroid, and major gaps are found at the values 3:1, 5:2, 7:3 and 2:1 (see Fig. 2.1).

How does this work? Why cannot asteroids stay in such orbits? Already Kirkwood must have wrestled with this question, and many did so after him, but it is only in the last decades that the picture

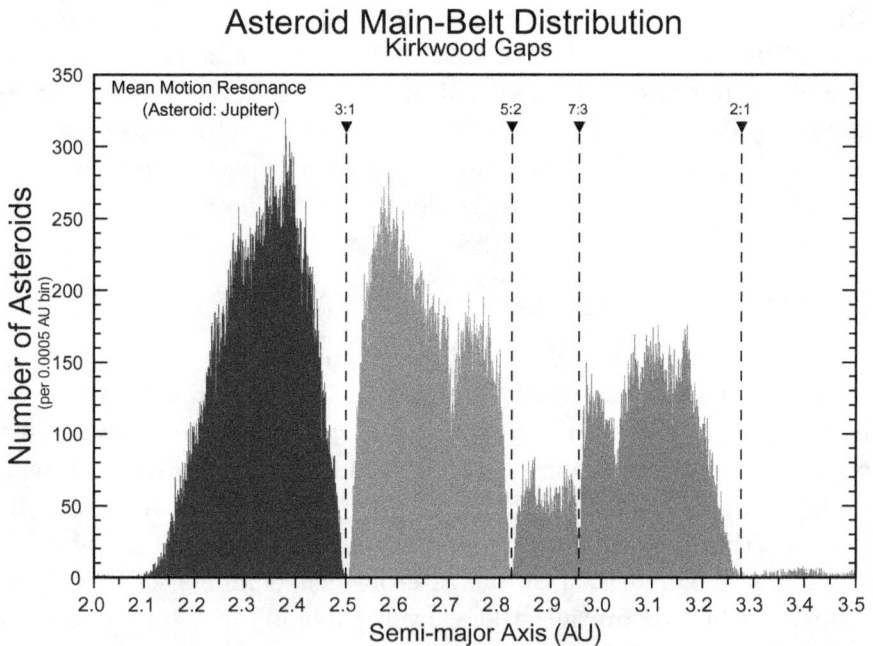

Fig. 2.1. The distribution of mean distances for minor planets in a range including the main belt. The uneven nature of this distribution is obvious. The main Kirkwood gaps are seen, and the corresponding resonances with Jupiter are marked as 3:1, 5:2, 7:3 and 2:1. Based on a plot by Alan Chamberlain. Credit: Alan Chamberlain, JPL/Caltech, NASA.

has become clear. Computer simulations and analytical theories have demonstrated that, within a million years at most, the resonant asteroid orbits attain so high eccentricities that they start crossing planetary orbits. When this happens, close encounters with the planet may cause the asteroid to change its orbit entirely and, above all, leave the resonance. Collision with the planet is another possibility, to which I shall return in Section 3.3.

Asteroids are not missing at the distances from the Sun, which correspond to the resonant mean distances, and this is due to the fact that the asteroid orbits are elliptic, spanning rather large intervals in real distance. Besides, there is also another kind of gaps, which show up if one marks the positions of the asteroid orbits in a diagram of orbital mean distances and inclinations. Once again, the reason is resonances, albeit of a different kind.

Asteroid orbits are perturbed by the planets — primarily by Jupiter. The orbit of an asteroid then turns in space. The ellipse turns in its own plane, and the plane turns relative to Jupiter's orbital plane. The eccentricity and inclination of the asteroid's orbit pulsate at the pace of these turnings. But the orbits of Jupiter and the other planets turn as well, because the planets perturb each other. In case an asteroid orbit turns at exactly the same rate as the orbit of a perturbing planet, a so-called secular resonance results. As an effect, the pulsating variation of the orbital eccentricity may run away to very large values. Then, once more, planetary orbits may be crossed and the asteroid may change its orbit at a close encounter or possibly collide with some planet. The most influential secular resonance in the asteroid belt ensues when the orbital plane of the asteroid turns at the same rate as the plane of Saturn's orbit.

In addition to the resonant gaps in the asteroid belt, there is also their opposite, i.e. concentrations, called families. The first of these were discovered a hundred years ago by Kiyotsugu Hirayama and are referred to as Hirayama families. However, the origin of the families has nothing to do with orbital perturbations or resonances. Instead, the family members are actually related. In fact, they are fragments of the same parent body — this will be further discussed in Section 2.4.

It is easy to realise that the total mass of the asteroid belt has to be much smaller than the mass of Earth, since the aggregate volume of millions of km-sized bodies is almost negligible compared to the volume of Earth. It is true that an important part is concentrated in the three largest ones (Ceres, Pallas and Vesta), but these are very far from providing the wanted mass. In fact, the total mass of the belt is estimated at about twice the mass of Earth's oceans — in other words, about one half per mille of Earth's mass. Somehow, the old Germans were actually right, because if the planets grew out of a giant disk of material surrounding the Sun, which was a popular theory as early as the early 19th century and still is a fundament of current theories (see Section 6.1), it would be unreasonable to think that this disk had an almost empty, broad gap between Mars and Jupiter. Even today, we imagine that the material of a planet must have been there from the outset, but in some way the construction of the planet failed and 99.9% of the material got lost. Similar waste may possibly exist in our world's big construction projects, if corruption is prevalent. What corrupted the building of a planet in the asteroid belt will be discussed in Section 6.2.

Of course, we are very interested to know what the asteroids consist of. But unfortunately, it is hard to get a detailed answer to the question. By observing asteroids spectroscopically and comparing with laboratory spectra of different minerals, we can reach a certain knowledge about what exists on their surfaces, but the question still remains about the composition at depth and, thus, of the bodies on the whole. On the one hand, the surfaces are influenced by the cosmic ray bombardment to which they have been exposed for an extremely long time, and on the other hand, asteroids might be chemically layered by melting, more or less as in the case of Earth.

As we shall see in Chapter 3, we have access to lots of samples of asteroids in the form of meteorites. Many of these are available at research labs for detailed analysis. One might therefore believe that the problem would be easy to solve, but this would require exact knowledge of which asteroids the particular meteorites originate from, and this is very rarely the case. Anyway, it has been possible to divide the asteroids into different classes, characterised by different surface materials. Largely, the occurrence rate of a given class is connected to the distance from the Sun, and the division into classes thus indicates a chemical stratification of the asteroid belt. To correctly understand what this tells us about the origin and evolution of asteroids is the main goal for much of the currently ongoing research.

The outer part of the asteroid belt is dominated by very dark and greyish bodies. This region also hosts most hints of water in asteroids. It seems evident that we are dealing with materials formed at low temperatures. The other kinds of asteroids, which dominate in the inner regions of the belt, are lighter and more reddish, and rather show signs of high temperatures. There is nothing strange about this if the heat comes from the Sun, but many details remain to be clarified. The two principal types mentioned are denoted by the letters C and S, respectively, because the dark and greyish hue of the former suggests the presence of carbon, and the lighter colour of the latter is believed to reveal common silicate minerals.

Far beyond the outer edge of the asteroid belt there is another very large group of minor planets, which has been known for over a hundred years. These are the so-called trojans, named after the Greek and Trojan war heroes according to the Iliad. Trojans have special orbits, much like that of Jupiter, and have, on average, the same period of revolution. They dwell in two wide regions situated around Jupiter's orbit 60° ahead of and 60° behind the planet. Those with Greek names move ahead of Jupiter, and those with Trojan names move behind.

As yet, only about 7,000 trojans are known, but an enormous number remain to be discovered. According to some assessments, the total number of trojans with a diameter of more than 1 km may be similar to that of the asteroid belt, although this may be an overestimation. In any case, the largest trojan, Hektor, is much smaller than the largest of the main belt, and the total mass of the trojans ought to be considerably smaller than the mass of the main belt. The observed trojans are as dark as the objects on the outskirts of the asteroid belt and sometimes even darker — as dark as comet nuclei. The relationships of trojans to main belt asteroids and comets is an interesting issue, to which I shall return in Section 6.3.

2.3. Transneptunians

Pluto was discovered in 1930 by Clyde Tombaugh. The next similar discovery took more than 60 years to occur, and during this whole time Pluto was regarded as a planet. Whether there was anything else in this region of space just beyond Neptune's orbit was a matter of speculation, but the issue did not arouse any immediate, lively discussion. Neither Pluto's mass nor its size was known with any considerable precision, and certainly not the mass. According to one estimate, Pluto would be the cause of perturbations on the orbital motion of Uranus, and its mass would then be comparable to that of Earth. This was, however, very far from the truth, and the perturbations were just measurement errors, but this could not be established at the time.

In any case, Pluto was the odd one out among the planets. It seems odd — nearly inexplicable — that the solar system would have four Earth-like planets, then four giant planets, and then yet another Earth-like planet. In addition, the orbit of Pluto is very improper for a planet. That is, it crosses the orbit of Neptune and only special conditions allow a collision to be avoided.

However, regardless of the true nature of Pluto, a query arose as to where the limits of the solar system are actually situated. It has always been hard to rule out the existence of yet another large planet far beyond the orbit of Neptune, but regardless of this issue, one may ask if it is reasonable to imagine the solar system to end abruptly with the giant planet Neptune. If the planets were built of smaller bodies — and this has long been a popular idea — it may be more natural to think of a gradual transition from the full-grown Neptune into the empty space between the stars via a region populated by smaller bodies, which never had the opportunity to assemble into a planet.

Such thoughts were brought forward in the early 1940s by Kenneth Edgeworth and at the beginning of the 1950s by

Gerard Kuiper. But there was no proof, and the ideas did not have any real impact. In most people's eyes, Pluto remained a mystical solitary at the edge of the solar system, the ninth and last discovered planet.

It would take until the 1980s before the situation started changing in earnest. An issue had been aroused concerning the origin of the short-period comets, i.e. those, which — similar to the previously mentioned Wild 2, Tempel 1, Hartley 2 and Churyumov–Gerasimenko — complete one revolution around the Sun in 5–6 years. My own doctoral work dealt with this topic, and I'll talk more about it in Chapter 4. But a few years into the 1980s it so happened that I heard about a new work by the young scientist Julio Fernández, who was from Uruguay but was active in Germany at the time (see Fig. 2.2). One of his results interested me in particular and made me rather distrustful. He argued that understanding where these comets came from required an explanation which was new to me at the time.

Fig. 2.2. Julio A. Fernández, Uruguayan astronomer who has made outstanding research contributions on the small bodies residing in or originating from the transneptunian region of the solar system. Courtesy J. A. Fernández.

What he was getting at was quite similar to what Edgeworth and Kuiper had written about much earlier, and even closer to the idea introduced by Fred Whipple — a kind of cometary belt just beyond Neptune's orbit. One day in the spring of 1983 I had lunch with my colleague Claes-Ingvar Lagerkvist, and he mentioned that we should be able to organise a meeting in Uppsala for our friends, acquaintances and other science contacts in Europe. I immediately joined in the idea, and we promptly started the preparations. The project involved people who conducted research on asteroids and comets, and we added meteors too by contacting our senior colleague at Lund Observatory, Bertil Anders Lindblad. The meeting was held in June 1983 and gathered almost 80 participants. Among the invitees was Julio Fernández himself, who gave a talk about his work with the Chinese native Wing-Huen Ip, concerning the very proposal on the origin of comets.

Alas, it was hard to convince me with regard to this issue, but anyway, Julio and I started a collaboration that would turn out important for both of us. In the meantime, a group of Canadians and Americans started to tackle the issue further by massive computer runs, and they came to results similar to those of Fernández and Ip. Now the whole research community reacted, and suitable telescopes were put in use to discover the proposed belt beyond Neptune, which had received the name Kuiper Belt, regardless of the fact that Edgeworth had been first.

As a sceptic, an interesting time came for me, too. Year after year passed without any discoveries, and the negative results began to be interpreted in terms of limits to the number of objects in the Kuiper Belt. But this trend was broken on August 30, 1992. This is when David Jewitt and Jane Luu discovered the object 1992 QB_1, which was later named Albion. Its diameter is about 100–150 km, and its orbit is situated about 44 astronomical units from the Sun with a very low inclination.

The 62 years of Pluto's hegemony in the transneptunian space came to an end. During the following years, Albion was followed by one new discovery after another, and the number of transneptunians rapidly soared. The situation reminded a bit of the reputed ketchup bottle. But it would take long before any new object challenged Pluto in size, and therefore the debate about Pluto's planethood took a considerable time to surface.

In the beginning, the observed transneptunians were of two kinds. The first kind had orbits rather similar to Albion's — roughly circular, and they were contained in the region between 40 and 50 astronomical units from the Sun; these made up the Kuiper Belt. The second type, with their orbital periods close to 1.5 times that of Neptune's, reminded of Pluto. This circumstance is one of the reasons why Pluto avoids colliding with Neptune, as is the case for the new objects that need such protection. Since all of them are much smaller than Pluto, they have been called plutinos.

But in 1996 it was time for a new category to be created. Unlike the Kuiper Belt and the plutinos, these objects enjoy no permanent protection from close encounters or collisions with Neptune. Some of them have orbits that even right now expose them to this risk. Others round the Sun at safe distances beyond Neptune's orbit, but over long enough times their orbits are perturbed, so that the risk arises repeatedly. Consequently, in the long run the objects are under Neptune's control and their orbits are unstable. They may sometimes receive energy from the planet, so that the orbits reach ever further from the Sun, and at other times they may lose such energy and instead become captured into the region between the giant planets. The objects orbiting beyond Neptune's orbit belong to something that, for lack of a better name, is called the scattered disk.

The masses of the Kuiper Belt and of the scattered disk are only crudely estimated. In both cases, the currently preferred value is about 1% of Earth's mass. This means 20 times more than the mass of the asteroid belt for each of the two populations. However, the uncertainty is large, especially in the case of the scattered disk, which reaches out to such large distances that even very big objects may be practically invisible to existing telescopes.

Minor planets moving between the orbits of Jupiter and Neptune have been known since 1977. The American astronomer Charles Kowal then discovered an object, which received the name Chiron. Its orbit has the inner turning point near the orbit of Saturn, and the outer turning point near Uranus's orbit. As mentioned, Chiron was the first, but we now know several hundred objects of a similar kind, and these have commonly been designated as centaurs. It is not likely that all centaurs have had a past existence in the scattered disk, but for most of them this ought to be the case. Centaurs are, therefore, usually considered a kind of transneptunians led astray, but as we shall see in Section 4.3, they are also an important link to the short-period comets.

As the scattered disk exploration progressed, knowledge about the structure of the Kuiper Belt also advanced. The most interesting discovery was that this belt consists of two distinct parts of different origins. One part may be called the native population, i.e. the bodies that were formed in the region where they are now residing, while the other part consists of immigrants from regions closer to the Sun. The two populations can be distinguished statistically by means of their colour and reflectivity, their propensity to be surrounded by satellites, and also their orbits around the Sun. These orbits have motivated the designations used.

The native population is called the cold Kuiper Belt because the orbits have very low inclinations and are nearly circular. The immigrants, on the other hand, have orbits that tend to be more inclined and are often more elliptic. In this sense they resemble the scattered disk, although with the important difference that they have been freed of the links to the planet Neptune. Their migration is thought to remind of the one that Pluto made (see Section 6.3), but the coupling to Neptune's orbit (in this case via the 2:1 resonance) was broken on the way, and instead of an external class of plutinos around 50 astronomical units, the solar system got the so-called hot Kuiper Belt.

The most recent new category of transneptunians was revealed by a remarkable discovery made in November 2003 by a team of astronomers headed by Mike Brown. They discovered the most remote minor planet of the solar system at a distance of 90 astronomical units from the Sun — three times as distant as Neptune. This was named Sedna, and its orbit is much more remarkable than the distance at which it was discovered. The minimum and maximum distances from the Sun are 76 and almost 1,000 astronomical units, respectively. The reason for the amazement of the scientists is that the inner turning point is so far beyond all the planet orbits. Sedna is therefore nearly unaffected by the planets, and it seems never to have been close to them. How then has it been placed into its curious orbit? I shall discuss this in Section 6.4.

Sedna is roughly the size of Ceres, but it was hard to discover due to the distance, and had it not been so close to its inner turning point, the discovery would have been impossible. Hence, we have to ask if there ought not to be even larger, undiscovered bodies on similar orbits at much larger distances? Yes, this should doubtless be the case. Seventeen years have passed since Sedna was discovered, and it is no longer unique in its category. Two more can be included, and these were discovered in 2012 and 2015. Both of them are smaller than Sedna and neither is currently more distant than Sedna.

There is not yet any official name for the category in question, since so few objects are known. Among proposed names we find "sednoids" and "Inner Oort Cloud". In any case, it is just the tip of an iceberg that we have seen. In fact, while Ceres is for sure the largest object in the asteroid belt, Sedna is no more than the currently largest object discovered in its category.

The same may likely be the case for the scattered disk. It hosts the as yet most massive among all known transneptunians — Eris. Concerning radius and volume, Eris is somewhat smaller than Pluto, but its mean density is 30% higher. It was discovered by the same team that discovered Sedna at about the same time, but the discovery of Eris was not announced until 2005. The distance upon discovery was larger than that of Sedna, but Eris is close to the outer turning point of its orbit, while Sedna is close to the inner turning point. Since there are plenty of objects in the scattered disk whose orbits reach much further out than Eris's orbit, we cannot know if Eris is really the largest. It appears that the largest object may still be hiding somewhere considerably further away.

Clearly, both the asteroid belt and the trojans pale into insignificance compared to the giant population dwelling beyond

Neptune's orbit. For myself, who grew up with the asteroid belt and who during his entire education did not hear a word about what might exist out there except Pluto, the discovery of the transneptunians was a shattering experience. This is certainly among the greatest of all the recent discoveries that have transformed our view of the solar system.

2.4. Collisions Among Minor Planets

The immense volume of space hosting the minor planets notwith-standing, there is, after all, some degree of congestion. Collisions happen now and then. This is largely due to the high age of the solar system. Usually, asteroids do not run any immediate risk of being hit, but when longer times are considered, the risk increases, and eventually they have to collect one punch after another like tired, defenceless boxers.

What is the consequence? It depends entirely on the size of the asteroid. In this case, it is good to be big. The relative velocity at which two asteroids collide is typically about 5 km/s, i.e. extremely supersonic. The result is an explosion that sends shock waves through both bodies in case they actually survive. This is far from the elastic knocks of the balls on a pool table. Asteroid collisions are not at all elastic, and the smaller object is often totally consumed by the explosion.

Consequently, an asteroid only lives until it collides with a partner big enough to destroy it. The bigger the asteroid, the fewer the fatal objects and the longer it survives in general. But as long as it lives, it collects the scars of crashes with considerably smaller objects. The smallest are the most abundant, and such crashes only produce small craters on the asteroid's surface. Indeed, all relatively large asteroids thus far explored are strewn with small craters. But on those that are old enough or strong enough, one can also spot the results of more dangerous collisions — craters so large that the survival of the asteroid may surprise. This was the case, for instance, with the asteroid Mathilde whose size is about 50 km, and which was visited by the NASA probe NEAR Shoemaker in 1997.

It is a rewarding idea that asteroids are approaching their demise. This makes you wonder, what traces they show of this fact, except for craters of various sizes? What else can mark the road to destruction? From the collision that creates the largest possible crater on an asteroid to the one that barely manages to destroy the asteroid completely, there is some leeway where different things can happen. One common outcome is that the asteroid is broken up into small pieces, which start to fly apart but are stopped by gravity and fall back together. The result of this process is called a rubble pile. In this case, the asteroid was close to destruction but just barely escaped.

As a matter of fact, most asteroids seem to be such rubble piles. The very largest ones have avoided the dangerous collisions, and the smallest ones have too weak gravities, but the rest are almost always rubble piles. There is also another closely related phenomenon of common occurrence — namely, asteroid satellites. These tiny moons are moving close to their mother asteroids and are likely fragments from major collisions, after which they ended up in satellite orbits instead of landing on the asteroid. Different methods have been employed to detect them, the first and most evident example coming from the space probe Galileo, which NASA sent out to explore Jupiter. In 1993, this probe passed close to the asteroid Ida, and the resulting images showed a small moon, which was named Dactyl. Ida has an oblong shape with a largest extension of somewhat more than 50 km, and Dactyl is roundish with a diameter of one and a half kilometre (see Fig. 2.3).

An interesting circumstance is that Ida is a Hirayama family member. It belongs to the Koronis family, named after its largest member. The family is made up of asteroids moving in individual orbits without any obvious connection. But after correction for the turnings and pulsations experienced by the orbits due to the planetary perturbations, the affinity of the asteroids emerges clearly. Basically, the orbits are very much alike, and they can be fitted together into a common origin, when the asteroids were sent out from a single point in space. The reason for this ought to be that a big asteroid — the mother of the entire family — was shattered by such a violent collision that its fragments were dispersed. This is

Fig. 2.3. Galileo image of the asteroid Ida, taken on August 28, 1993. The small dot to the right is Ida's satellite, Dactyl. Credit: NASA/JPL/Processed by Kevin M. Gill.

verified by the observations of the Koronis family, since its members fit together geologically as well.

Thus, the families provide direct evidence for the demise of asteroids. One may even surmise that almost all main belt asteroids are fragments of shattered mothers at the same time as they themselves are slowly but surely approaching their destruction. From this perspective, the asteroid belt appears like a crusher of the solar system. Do we here see the answer to the question, what prevented the construction of the "intended" planet in the asteroid belt in the youth of the solar system? The response is essentially no, even though it is clear that the comminution caused by collisions has had a certain influence. That this can be said is due to a circumstance to be described in the following section.

Now, what about the collisions in the Kuiper Belt? If we argue based on the number of transneptunians currently populating the Kuiper Belt, we find that the collisions there are much less important than in the asteroid belt. The main reasons are the enormous volume that the bodies are residing in and the lower velocities at which they are moving. But at the same time, we find the opposite message. In the asteroid belt, the largest bodies have escaped catastrophic collisions, but in the Kuiper Belt the situation is the opposite. The two largest ones are called Haumea and Makemake, and both are considerably larger than Ceres. But contrary to Ceres, both have satellites and, in addition, Haumea is the largest member of a family where several members have diameters of hundreds of kilometres.

This could not be the case unless the members of the Kuiper Belt had earlier been moving in a much more crowded and violent environment. Moreover, Pluto, with its special orbit, also has a very large satellite, whose origin can most easily be understood if we assume that it was separated from Pluto as a result of a giant impact. What is more, Eris — the largest member of the scattered disk — has a satellite of considerable size, too. Did they all once crowd in the same, much less spacious environment? Here, we are glimpsing one of the greatest discoveries concerning the solar system in recent times, which I shall account for in Chapter 6.

2.5. Large Asteroids

Let us now visit a number of minor planets (and dwarf planets), all of which have been visited by space probes. To these we add a few others, which are awaiting such visits, but about which we still have learnt important things. We take them in order of increasing distance from the Sun.

Vesta is the second largest asteroid. With a mean diameter of 525 km, it is almost negligibly larger than Pallas, which is only 10 km smaller. Vesta's orbit is mostly contained in the inner part of the asteroid belt. It has long been known that Vesta is peculiar, when it comes to the composition of its surface layer. Its spectrum differs markedly from those of other large asteroids. An interesting discovery was that Vesta's spectrum is shared by a special kind of meteorites — the HED meteorites, which will be described in Chapter 3. The similarity between Vesta and the HED meteorites and their dissimilarity to other bodies are so large that there must be a relationship. Since then, it has been strongly suspected that the HED meteorites originate from Vesta's surface layer.

Even more interesting is the analysis of HED meteorites, which shows that they have a volcanic origin, unlike the vast majority of meteorites coming from the asteroid belt. Thus, it is a question of rocks that have been molten and re-solidified in the manner that continuously takes place in Earth's crust. The conclusion is that Vesta has a crust as well, and, fundamentally, that Vesta is like a miniature of our globe — a feature that apparently renders it unique in the asteroid belt of our days.

An additional clue emerged with the discovery of the Vesta family. This is a rather small asteroid family, where all the members share the same spectrum as Vesta. Its origin can be traced back to a collision, where Vesta lost a fraction of its surface layers, and part of the fragments now form the Vesta family. Moreover, some of these have had their orbits stretched out by Jupiter's perturbations, so that they might in principle collide with Earth. Fortunately, we have not seen this happen, but they themselves have been bombarded so that small splinters have been knocked off. Some of these splinters are found here on Earth as HED meteorites.

In 2007, NASA launched the space probe Dawn in order to explore both Vesta and Ceres. The probe arrived at Vesta in July 2011 and stayed in orbit for more than a year before it was sent further to Ceres. As expected, the images of Vesta show a surface strewn by craters, whereof most are small but some exceed 50 km in diameter. Even larger craters appear to be hidden under newer and smaller craters. But this mainly concerns Vesta's northern hemisphere. Around the South Pole there is something different, which exceeds the expectations of the scientists. One finds relatively few craters, but there are two — partly overlapping — giant impact basins instead (see Fig. 2.4).

The older basin, named Veneneia, has a diameter of 400 km and is partly covered by the younger Rheasilvia, which measures as much as 500 km. The images show Rheasilvia to be a very impressive natural formation. This basin has a depth of up to 20 km in relation to the surrounding area, and in the middle a central peak rises to a height of more than 20,000 meters. Of all the mountains in the solar system, only Olympus Mons on Mars is higher.

When was this feature formed? Unfortunately, there is no obvious answer to this question. The only clock at our disposal is provided by the Vesta family members, because their size distribution suggests

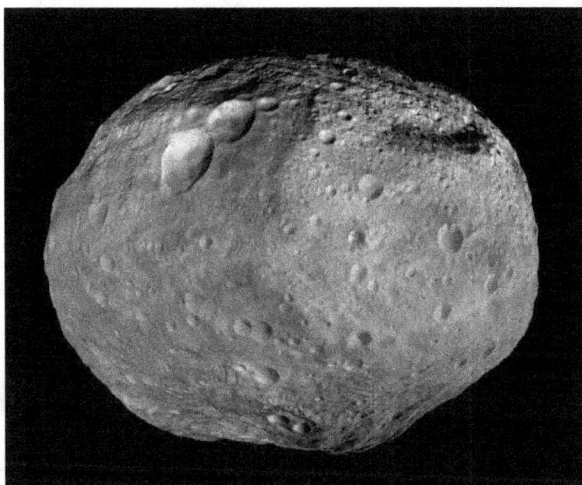

Fig. 2.4. The second largest asteroid, Vesta, imaged by the Dawn spacecraft. On the lower (south) side of the asteroid, the Rheasilvia basin and its central mountain are seen. Credit: NASA/JPL-Caltech/UCAL/MPS/DLR/IDA.

the age of the family at about 1 billion years. This might also be the age of Rheasilvia, since the basin is large enough to contain all the family members with a margin. A collision of this magnitude would have destroyed essentially every other asteroid, and even Vesta must have cracked but survived thanks to its size.

Concerning Vesta's history, incidentally, we know that the crust is truly ancient through dating of the HED meteorites (see Section 3.3). Vesta was formed within a few million years after the birth of the solar system, and the volcanism that created the crust started at once. Hence, we also know that this 20 km thick layer has survived the entire period of more than 4.5 billion years that has elapsed since then. The Rheasilvia collision was likely the most perilous threat, but no impact has been able to break up or crush the whole crust. From this, we learn the following. In case nearly the whole original asteroid belt would have disappeared through comminution due to collisions, it would be unbelievable that Vesta's crust survived. Hence, most asteroids must have vanished in some other way (see Section 6.2).

The next halt on our journey is Lutetia. This asteroid was discovered as early as 1852 by the Parisian Hermann Goldschmidt from his own balcony, and Lutetia is indeed the name for Paris used by the ancient Romans. It is truly a large asteroid with a mean diameter of about 100 km. Its orbit is situated in the central part of the asteroid belt. It was selected as a target for the ESA Rosetta probe during its last passage of the asteroid belt before the appointment with its real goal, comet Churyumov–Gerasimenko. In July 2010, Rosetta passed at a distance of only 3,000 km from Lutetia, and the images revealed an irregularly shaped, angular and cratered body.

In the camera team, we were initially surprised by Lutetia's high mean density, $3.4\,\mathrm{g/cm^3}$. We had certain ideas about which type of meteorites would best represent the material of Lutetia, but these had too low density. However, in retrospect, I do not see anything strange in this conflict. We should perhaps have chosen a different type of meteorites for the comparison. Furthermore, Lutetia is more compact (less porous) than we imagined, and lastly, an erosion of the surface layer may have played a role. The year after the Rosetta observations of Lutetia, Dawn arrived at Vesta, and Vesta's mean density was found to be very close to Lutetia's. It is therefore of interest to compare these two large asteroids.

There is an important difference in their geological evolution. As noted, Vesta is like an Earth globe in miniature. Its iron core is more than 200 km in diameter, and it has a volcanically formed crust. Lutetia has none of these. But due to the heating that it once experienced, some degree of melting may have occurred so that iron and nickel to some extent left the layers close to the surface. Thereafter, the surface has been eroded by impacts, and Lutetia has then preferentially lost its lighter constituents. Something similar must have happened, when the Veneneia and Rheasilvia basins were formed on Vesta.

The largest crater on Lutetia is called Massalia and measures 55 km in diameter. This impact, though, did not destroy Lutetia, which seems compact and coherent rather than a loosely configured rubble pile. A related finding is Lutetia's high age, which can be inferred from the very large number of craters on parts of its surface. The result is about 3.6 billion years, but the age may in fact be

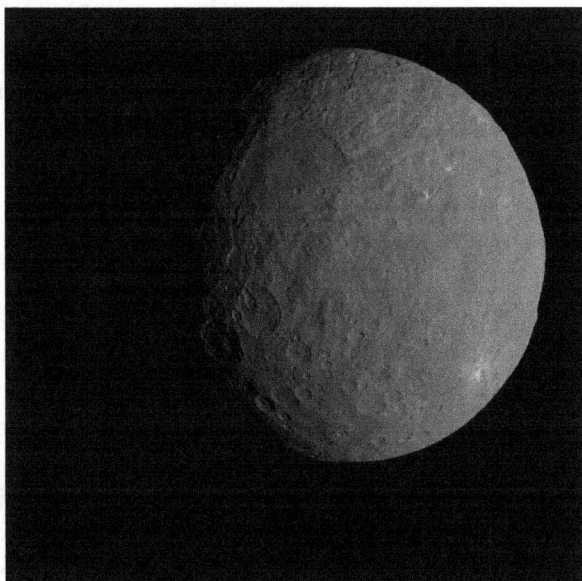

Fig. 2.5. The largest asteroid, Ceres, imaged by the Dawn spacecraft. Credit: NASA/JPL-Caltech/UCAL/MPS/DLR/IDA/Justin Cowart.

even higher. Therefore, Lutetia is considered to be one of the few survivors in today's asteroid belt — a testimony of the bodies that originally populated the region between Mars and Jupiter.

In March 2015, the Dawn probe arrived at Ceres (see Fig. 2.5). It then faced an asteroid and a dwarf planet, which is very dissimilar to Vesta. Above all, Ceres consists of different materials. This is revealed directly by the difference between the spectra of the two objects, as far as the surface material is concerned. It has been known for a fairly long time that there are large differences also at depth: estimating the masses and volumes of both objects showed that Ceres has a much lower mean density. Dawn contributed to an improved understanding of Ceres's constitution, partly through detailed studies of the surface layer, and partly by charting the gravity field of the dwarf planet, through which we have information about the variation of the density with depth.

Ceres's mean density of $2.16 \, \text{g/cm}^3$ is so much lower than the densities of all reasonable rocky materials that it must contain significant quantities of ice or H_2O in some other form. In the latter case, it could be a question of liquid water or water bound up in hydrated minerals. The big issue, which not even Dawn has been able to answer, is how large a fraction of the mass of Ceres is made up of H_2O. The orbit of Ceres is confined to the outer part of the asteroid belt, though not very close to the outer edge. The topic of debate has been whether the abundance of ice in Ceres is typical of this region or not.

It has long been evident that Ceres has a somewhat peculiar spectrum, compared to most of its neighbours, but this tells nothing about differences in abundance of ice. One has rather concentrated on a sort of meteorites that is considered to come from asteroids in the outer part of the belt — the so-called carbonaceous chondrites. Some of these contain up to 10% of water in the form of hydrated minerals. How well does this fit with the abundance of H_2O in Ceres? In case it does not fit at all, one may speculate that Ceres is a stranger that has migrated into the asteroid belt from an altogether different place. This might, in turn, bring information about what happened when the solar system was formed.

As for myself, I prefer to take this with a grain of salt. According to the gravity data from Dawn, Ceres is composed of a core of hydrated rocks, typical of carbonaceous chondrites, plus a mantle, which largely consists of ice. The thickness of the mantle is estimated at 100 km, while Ceres's radius is 470 km. This would signify a global H_2O content larger than that of the average asteroids in Ceres's vicinity. But, perhaps, we see the result of collisions instead of a genuine, distinctive feature of Ceres.

The separation of ice and rocks inside Ceres is far from complete but still prominent, and this must have come about very early in connection with the formation of Ceres. A similar separation also occurred inside Ceres's smaller neighbours. Much later, as the collisions set in, bodies of all different sizes with icy mantles and rocky cores were affected. As a result, icy mantles have only survived on Ceres and some others among the largest bodies. These are the only survivors in the outer part of the asteroid belt, while all the other bodies are fragments — some perhaps icy but most of them probably rocky. From the latter, we get carbonaceous chondrites as meteorites, while the icy small fragments have boiled away en route to Earth and in any case did not penetrate to the ground.

The presence of ice in other large asteroids moving in the outer part of the belt is verified by observations. In the first place, this concerns the asteroid Themis with a 200 km diameter, whose surface is more ice-rich than that of Ceres. Furthermore, Themis is the largest member and namesake of its own Hirayama family, in which other members too are ice-covered. Two of these have been observed as comets when passing closest to the Sun because of an intense flow of water vapour from their sublimating surfaces. They are counted with a group of objects called main belt comets.

The Dawn images of Ceres also show that the surface is rich in craters, but that there is a dearth of large craters. The most likely reason is geological activity. Large craters also tend to be deep, and on Ceres they may have reached so deep that water and ice has flown up and filled them. A contact between the surface and the interior is also suspected due to a number of light spots. In general, the surface of Ceres is dark and carbonaceous, but in these spots, it appears that saline solutions have extruded via cracks and pores in the icy mantle. As the water evaporated, the salts crystallised and got trapped on the ground. A few examples are sodium carbonate, ammonium chloride and ammonium carbonate.

2.6. Large Transneptunians

We now move to the transneptunians, but let us stop for a while among the large satellites of the giant planets. In fact, some of these are captured minor planets. In recent years, many relatively small moons have been discovered in remote orbits around all four giant planets. It is generally agreed that the planets captured them during a chaotic epoch during the early history of the solar system. Thus far, though, we know very little about their properties. The exception is Saturn's satellite Phoebe with a diameter of more than 200 km, which was passed at close range by the space probe Cassini in 2004. The images of Phoebe show a fairly round celestial body, which exhibits

Fig. 2.6. Saturn's large and distant satellite, Phoebe, imaged by the Cassini probe on June 11, 2004. The picture shows Phoebe at first quarter phase similar to the Halfmoon, which makes the shadows cast by craters and basins stand out clearly near the terminator. Credit: NASA/JPL/Space Science Institute.

lots of craters and gives the impression of being nearly destroyed by collisions (see Fig. 2.6).

The surface is dark, but much lighter material appears inside the craters. This is ice, which, as far as we can judge, contributes a large fraction of Phoebe's mass. Indeed, the density is only $1.6\,\mathrm{g/cm}^3$.

However, the largest captured satellite is believed to be Neptune's satellite Triton. Just like Phoebe, Triton is thought to be an original cousin of Pluto, Eris and all the comets — more about this will be discussed in Section 6.3 (see Fig. 2.7).

With a diameter of 2,700 km, Triton is a little larger than Pluto, and these two are accidentally quite similar to each other. How was Triton captured? There are two proposed mechanisms that may work, and I don't dare say which one it was in reality. In any case, this happened early in the history of the solar system, at a time when Neptune's orbit migrated outward and gradually

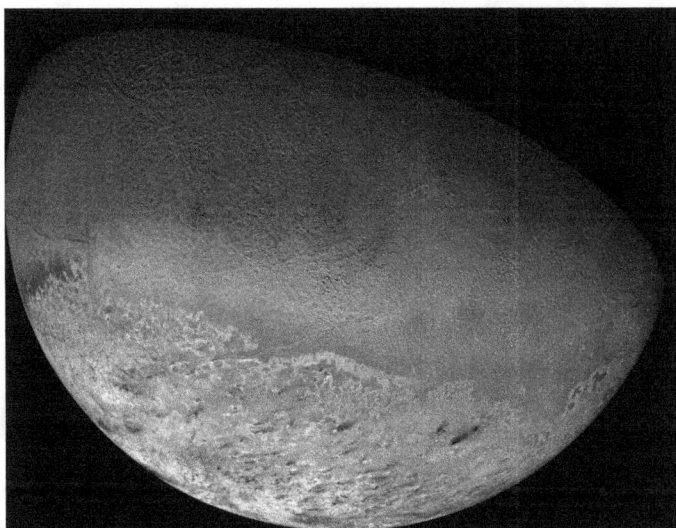

Fig. 2.7. Neptune's largest satellite, Triton, is likely a captured, originally transneptunian object. This picture was acquired by the Voyager 2 spacecraft during its flyby of the Neptune system in 1989. The surface is generally dominated by nitrogen frost, but on the lower (south) side a polar cap is thought to contain methane ice. Credit: NASA/JPL/USGS.

grew wider. The planet then happened to approach the minor planets that had been born out there and did not participate in the planet's migration. Triton was among those that came closest to Neptune and experienced the strong gravity of the planet.

There are now two possibilities. Either Triton happened to collide with one of Neptune's moons so that it was slowed down and itself became a new moon, or Triton itself had a moon that it lost due to Neptune's influence. If so, this loss might decelerate Triton into satellite capture around Neptune. Regardless of what really happened, Triton is interesting because it was born in freedom as one of the truly large objects in the region of space beyond the giant planets and only later became demoted to a simple Neptunian satellite.

Demotion is a word that was actually used in the debate around Pluto's status as a planet before the decision at the IAU General Assembly in Prague in 2006. It may have had some significance that

Fig. 2.8. Close-up image of Pluto, taken by the New Horizons spacecraft on July 14, 2015. The large contrast between dark and bright terrains is noteworthy, but the overall reflectivity is very high. Credit: NASA/JHUAPL/SWRI/Alex Parker.

an American space probe already was on its way to Pluto to explore what was then the ninth planet of the solar system and the only one that had never seen such a visit. The name of the probe was New Horizons, and it arrived at Pluto on July 14, 2015. It really does not matter which label we stick to this fantastic object, because both the images and other results are in any case extremely interesting (see Fig. 2.8).

On Pluto's surface just like Triton's, all kinds of rocky material are completely absent. The surface is ice-covered, albeit not by the usual H_2O ice, but mostly by frozen nitrogen. Also, frozen carbon monoxide and frozen methane exist there. From the fact that such extremely volatile substances are frozen into ice, we realise that the temperature is very low. The Sun shines very weakly from a large distance, and the icy ground reflects away most of the solar energy. As if this were not enough, the ground is sweating off the measly energy received, since the ice sublimates so that the surface is further cooled. A thermometer on Pluto would show about 40 degrees above absolute zero (about $-230°C$) if the Sun is up and even less during nighttime. At some places, mountains of water ice are rising to heights of thousands of meters. Two of these regions, Tenzing Montes and Hillary Montes, are named after the two men who, in 1953, were the first to climb the peak of Mt Everest: the sherpa Tenzing Norgay and the New Zeelander Edmund Hillary.

New Horizons could only image part of Pluto's surface. The flyby was fast, and Pluto did not have time to rotate considerably with its period as long as 6.4 days. Thus, the whole night side was invisible, and moreover, the day side was partly on the side of Pluto that was turned away from the trajectory of the spacecraft. Of what is, after all, seen, there are two things that catch attention: the very bright Tombaugh region (named after Pluto's discoverer), and the — just as extremely — dark Cthulhu Macula (named after a figure in an American horror novel). Due to their shapes, they are also often called the "heart" and the "whale", respectively. In the western part of the heart, called Sputnik Planitia, the icy surface is extremely flat. Not a single crater is seen despite the fact that other regions on Pluto exhibit a huge number of craters. Obviously, something radical has happened here just a few hundred thousand years ago. The ice may have welled up in many places and sunk back elsewhere. Thus, the ground is broken up into different blocks, and this may still be going on. However, the reason remains unknown.

Our knowledge of Pluto's inner constitution is quite limited. Its low density of $1.85\,\text{g/cm}^3$ (somewhat lower than that of Triton, which is $2.06\,\text{g/cm}^3$) is reminiscent of Ceres, and Pluto too is largely composed of ice. One tends to presume that the internal heat has separated ice and rocks and to guess that the rocky core spans 70% of the radius, while the remainder of 30% consists of an icy mantle. One characteristic property of Pluto is that it has a giant moon, Charon. With a diameter of half that of Pluto (1,212 km against 2,375 km) and more than one tenth of Pluto's mass, Charon may be seen as a serious candidate for supremacy in the system. This also comprises four smaller satellites, whose orbits embrace both Pluto and Charon.

With a density of $1.70\,\mathrm{g/cm^3}$, Charon is fairly similar to Pluto in terms of composition, but it is not known whether the available heat has been enough to divide it into an icy mantle and a rocky core. There are no geologic formations such as Sputnik Planitia, but there are indications that the oldest craters were wiped out by geologic activity on some occasion long ago. An important issue deals with the origin of the binary system formed by Pluto and Charon. The most popular theory involves a giant impact of the same kind as that which created the Earth–Moon system. But this theory has its problems, and there is also another possibility. Namely, that the two bodies were separated already when they were formed in the infancy of the solar system out of a common, collapsing particle cloud (see Section 6.2) as a consequence of the cloud rotating too rapidly.

The surface of Charon lacks the types of ice that dominate on Pluto's surface. Instead, usual H_2O ice dominates there, but the North Pole is covered by a dark, orange-brown cap, which consists of a different material. In the 1970s, the famous Carl Sagan and his collaborator Bishun Khare performed experiments where a gas mixture containing, for instance, methane, ethane, ammonia and water was irradiated by ultraviolet light and bombarded with lightnings. In this way they produced a brownish, polymeric substance, which they called "tholin". Their work has had a large impact, and tholin has been considered to contribute to the surface material on many solar system bodies as well as to the haze in certain atmospheres. Indeed, the simple compounds that Sagan and Khare used are present here, there, and everywhere — primarily in the outer parts of the solar system — and both UV light and cosmic radiation are also omnipresent. However, one should consider that tholin does not have a given chemical formula. Different kinds of tholin can exist at different places, depending on which substances were there from the outset. One therefore often uses the plural term tholins.

Charon's polar cap is thought to be an example. Gases can reach the surface of Charon via Pluto's atmosphere and be deposited as frost during the long polar night of Charon. When dawn breaks, the sunlight creates small amounts of tholin, which remain as the rest of the frost evaporates during the day. As time goes on, this may build up tholin caps at Charon's poles. It has also been speculated that Cthulhu Macula and other dark regions close to Pluto's equator would contain tholin.

Quite naturally, it is only the largest bodies in the Kuiper Belt that have been targets of in-depth observations. At least one of these, Makemake, is actually visible through advanced amateur telescopes every year in the month of March. Tombaugh could have detected it, too, if he had pointed the telescope in Makemake's direction. This nurses yet another thought about Pluto's special place among the transneptunians. If we accept that Triton was born free and is truly a transneptunian like the others, it may be counted as both the largest and the first discovered. But if it had not been shining close to Neptune thanks to its satellite capture, William Lassell would not have seen it in 1846. It is impossible to say what would then have happened to it and if it would have been discovered at all.

Makemake and Haumea are about the same size, and both are situated a little more than 50 astronomical units from the Sun. Since Makemake's surface has a higher reflectivity, it shines brighter of the two. The dominant surface material is methane ice, but due to an orange hue, tholin is also assumed to exist there in large quantities.

The issue of the sizes of the bodies is worth a comment. There are serious difficulties in determining the size of a transneptunian from its visible and infrared radiation. Even from the foremost, satellite-based telescopes, the results have therefore shown large scatter. The most reliable method, however, is built on the fact that the light from a star can be blocked when a minor planet passes across the line of sight — this is called an occultation. Such measurements have yielded an estimated diameter of somewhat more than 1,400 km for Makemake, whose shape seems rather round. The density is estimated at somewhat less than $2\,\text{g/cm}^3$.

The uncertainty is larger for Haumea, and its shape is strongly elongated. According to some estimates, its longest axis is shorter than the diameter of Makemake. If so, its volume is considerably smaller than that of Makemake, and the density would be as high as $3\,\text{g/cm}^3$. But an observed occultation in January 2017 yielded a new size estimate. The longest axis would now be comparable to Pluto's diameter, the volume would be much larger than that of Makemake, and the density would be only about $1.8\,\text{g/cm}^3$.

It is too early to tell what the truth is, but it is clear that Haumea has been the victim of a violent collision, since it hosts an important collisional family. Haumea's two moons are further witnesses. If it were separated into a rocky core and an icy mantle before the impact, much of the icy mantle would have been peeled off. The other family members would consist of this material. A high density of Haumea would then be likely.

In any case, the surface of Haumea is covered with ice. It has also been found that the ice is largely crystalline, similar to the ice we are used to on Earth. However, this is not to be expected for a transneptunian, which is permanently bombarded by cosmic rays. The crystal structure of the ice should then be destroyed fairly rapidly. That this has not happened on Haumea is interpreted to mean that Haumea's surface for some reason has recently been re-created. Smaller impacts may have played a role and also helped keeping the surface clear of extremely volatile substances like methane — Haumea's spectrum shows no signs of those.

Eris is currently at a much larger distance than the mentioned bodies. Its diameter was initially considered to be larger than that of Pluto based on its radiation, but an occultation in 2011 gave a new value of about $2{,}330\,\text{km}$, i.e. 2% less than Pluto's. This implies an extremely high reflectivity of 96%. Hence, the surface must consist of pure ice, which seems to consist mostly of methane. No particular hue has been identified, and this means that the tholins that must be produced there are buried beneath newly formed methane frost. They may possibly peep out more than 500 years from now, when Eris approaches the inner turning point of its orbit and the frost evaporates away.

Eris has a rather large moon. It has been named Dysnomia and its diameter is about $700\,\text{km}$ — more than Vesta's and Pallas's. Based on its orbital motion, we have our estimate of Eris's mass, which results in a remarkably high density of $2.5\,\text{g/cm}^3$. Either Eris was formed having a relatively small fraction of ice, or large collisions peeled off a significant part of the icy mantle long ago. Dysnomia's dark colour with a reflectivity of only 4% is also remarkable. Probably, the collision from which Dysnomia originates led to the loss of all substances more volatile than water, and the surface was covered by dark, carbonaceous material.

The surface of Sedna is considerably darker than the surfaces of Haumea and Makemake, and is extremely red as well. Large quantities of tholins are supposed to be present there, causing this hue. Ices of methane and methanol have also been traced in Sedna's spectrum. Collisions are rare on this remote body, and its surface is therefore believed to have been left in peace and quiet since a very long time ago. The diameter is about 1,000 km, and no satellite has so far been detected.

Let us now accompany the space probe New Horizons, which more than three years after the flyby of Pluto and Charon encountered a rather small and anonymous member of the Kuiper Belt at a distance of 44 astronomical units from the Sun. It was indeed nameless, since it was discovered as recently as 2014. But it was accidentally close to the orbit of the probe and thus was duly imaged, and thanks to the fine pictures, it has become one of the most illustrious minor planets of recent years. It was officially named Arrokoth, which alludes to a word of an American native language from the region around Maryland. Before then, the nickname Ultima Thule was used, since this is the most distant object that has ever been visited by a spacecraft. It is also clearly binary in a way that reminds of the nucleus of comet Churyumov–Gerasimenko. The two parts were thus called Ultima and Thule, respectively.

Arrokoth really looks like a snowman. It is true that the colour is too red, but the shape cannot be mistaken. Its origin is evidently explained by what I called the snowman theory in Section 1.6. The head and the body were connected by a very soft and gentle collision during the infancy of the solar system. Looking closer at Arrokoth's surface, one is struck by the dearth of craters. Other ancient bodies tend to be completely strewn, but this is not the case here. I think that this is just the way it should be. Arrokoth is a member of the cold Kuiper Belt, as shown both by its colour and its orbit. As such, it was never exposed to the inferno of crashes, which ravaged the bodies closer to the Sun and which destroyed Pluto. It rather exhibits the tranquillity that has always prevailed on the outskirts of the solar system.

2.7. Three Little Minor Planets

As we have seen, some minor planets are very large, but after all, the vast majority only measure a few kilometres or less. These are normally difficult to observe, since the large distance renders them very faint. But, in recent years, three asteroids with diameters less than 1 km have had extremely close spacecraft visits. All three belong to the category of asteroids, whose orbits cross Earth's orbit and which are called Apollo asteroids. Their names are Itokawa, Ryugu and Bennu in chronological order of the visits.

The Japanese space agency JAXA has shown particular interest in this research. The JAXA probe Hayabusa arrived at Itokawa as early as 2005. It delivered close-up images of a curious object, whose shape reminds of an otter. The length is more than 600 meters and the width is 250 meters. However, its most remarkable feat was a low sniffing manoeuvre, during which a small sample of tiny grains was collected from the surface and brought back to Earth for analysis.

From the determination of Itokawa's mass, it has been realised that the asteroid has a very low density $(1.9\,\text{g}/\text{cm}^3)$ and hence a high porosity. Indeed, the material of this S-type asteroid is composed of silicate minerals with much higher densities. Itokawa is obviously a rubble pile without significant strength. This seems to explain a remarkable dearth of impact craters, since the unavoidable collisions shake the whole asteroid so that the scars of earlier impacts are rubbed out. Large parts of Itokawa's surface are strewn by big boulders, while others appear flat and even.

Chemical analysis of the grains has shown that Itokawa's stony material is most similar to a kind of meteorites denoted LL chondrites (see Section 3.3). Small amounts of crystal water can sometimes be found in such meteorites, and the same goes for Itokawa.

The hydrogen of this water has been found to contain about the same fraction of deuterium (heavy hydrogen) as the water in Earth's oceans. But even though this is the case, do not believe that Earth's water arrived with Itokawa-type asteroids! They are far too unusual now, and certainly also were at the time when Earth was formed. The origin of our water will be discussed in Section 7.1.

In 2018, the next visit of a small asteroid by a Japanese spacecraft took place. The probe was called Hayabusa 2, and the target is named Ryugu. This time a different kind of asteroid than Itokawa was selected. Ryugu is a C-type asteroid, which renders it extra interesting. Indeed, the asteroids are seen as possible witnesses of the formation of planets, and the more geologically unaffected they are, the more confidence we have in their testimonies. A history at high temperatures is thought to indicate extensive modification, and the asteroids that have escaped this fate are thus considered extra valuable. The richness of C-type asteroids in carbonaceous and water-bearing minerals gives them such advantage, since it appears to indicate formation and evolution at low temperatures (see Section 2.2).

Ryugu's diameter is more than 800 meters, and the shape is fairly round. Like Itokawa, it has a remarkably low density ($1.2\,\mathrm{g/cm^3}$) and must hence be highly porous. The close-up images from Hayabusa 2 show its surface to be strewn with large blocks and boulders similar to what was seen on Itokawa. Two small, shattered rubble piles — this seems to be what the Japanese probes have visited. But Hayabusa 2 had a decidedly higher level of ambition than Hayabusa. Four small rovers were landed on Ryugu's surface, and samples were acquired at several places. At the time of writing, the probe is on its way back to Earth to deliver its cargo.

Soon after Ryugu was visited, an American space probe OSIRIS-REx arrived at another small, Earth-crossing, C-type asteroid named Bennu. This probe, too, caught surface samples to be brought back to Earth for analysis in a few years. Bennu and Ryugu have exhibited considerable similarities, even though Bennu is smaller, with about 500 meters in diameter, and its surface is even more covered by large blocks than those of Ryugu and Itokawa.

I finally want to mention that Bennu has enjoyed a certain attention in news media, because it is currently the most "dangerous" of all asteroids. The probability of it hitting Earth in the future is very small, and it will take a long time until the danger arises, but the risk is indeed larger than in other comparable cases. I shall write more about this in Section 5.4.

Chapter 3

The Smallest Pieces:
Meteors and Meteorites

How small can a minor planet be? The question may seem odd but can still be answered. The concept of "minor planet" stands for an astronomical object, and as such it must be observable by telescopes. Nowadays, one may often read about asteroids passing close to Earth, which are no larger than small lorries. These are the smallest asteroids that we know of, but they are, of course, special by having orbits that cross the orbit of Earth. Bodies of similar sizes must exist in large numbers in the asteroid belt, but these are beyond reach of our telescopes. Yet, we can, of course, speak of them, and it is then legitimate to use the word "asteroid". But it is conceivable that the telescopes of the future will be able to reach even smaller objects as they pass close to Earth. In principle then, a minor planet might be arbitrarily small, because space in the solar system is populated by dust grains, pebbles and rocks of all possible sizes.

But cross that bridge when you come to it. For the time being, all those things that the telescopes cannot see but that keep hitting Earth, can be classified under a category of meteoroids. This strange word has a logical explanation. In fact, regardless of the size — from a fraction of a millimetre to several metres — they cause the phenomenon called meteors when they penetrate into Earth's atmosphere. The word meteor is derived from the Greek *meteoros*, which means "up in the air", and it is thus related to the word meteorology. There is yet another term, which is often confused with the others, and this is meteorite. When one sees a meteor, it is always a meteoroid that enters the atmosphere, and that almost always evaporates in high atmospheric layers. But in very rare cases, parts of the meteoroid manage to penetrate all the way to the ground and may be picked up. We then speak of meteorites.

Every meteoroid in the solar system follows an orbit around the Sun. In general, these orbits are situated close to the plane, to which the planetary orbits are concentrated (the ecliptic plane). The entire population of meteoroids, thus, forms a flattened, disk-like cloud, which encloses the planetary system. We are situated inside this cloud and are sometimes able to see it at sunrise or sunset as a faint, diffuse light along the constellations of the zodiac (see Fig. 3.1). This is known as the zodiacal light, but unfortunately, very few people enjoy the advantage of a dark enough sky to actually see it. Many space probes have brought along instruments to study the structures of this cloud of meteoroids *in situ*, but we will now concentrate instead on those that are studied with Earth itself as a detector, i.e. meteors and meteorites.

3.1. Shooting Stars and Fireballs

The escape velocity from Earth is 11 km/s, and it does not matter much if we count from the ground or from the highest atmospheric layers. When small particles hit Earth, they often have initial velocities that are even larger — maybe comparable to Earth's own velocity of 30 km/s in its orbit around the Sun. To that we have to add the extra velocity that Earth's gravitational acceleration implies, which would be 11 km/s if the initial velocity were zero. As seen, the velocities of meteoroids entering the atmosphere are almost unimaginable.

Let us imagine that we follow a centimetre-sized meteoroid along its path through the atmosphere. In the beginning, the air is so thin that the meteoroid cannot feel it but just collides with one molecule after the other, and the effect of this is negligible. At a height of about 100 km, the air is thick enough to change the situation fundamentally. The meteoroid now has to plough its way through a coherent gas.

Fig. 3.1. The zodiacal light, observed from Cerro Paranal in northern Chile, where the ESO Very Large Telescope is located. Courtesy: ESO/Y. Beletsky. License: Creative Commons Attribution 4.0 International.

Due to its extremely supersonic speed, a strong bow shock is formed ahead of it. The air inside the shock front, which surrounds the meteoroid, is strongly condensed and incredibly hot. It shines so brightly that we can see it from the ground, and the meteoroid is burned up. This process only lasts for a second or so and takes place at a height of about 80 km. What I have described is usually called a shooting star, and for astronomers it is a meteor.

If the particle is much smaller, the air does not light up, but radio waves are still reflected by the shock front, and it can be observed by radar. On the other hand, if the particle is much bigger, the meteor shines much more brightly, penetrates much deeper into the atmosphere and lasts much longer. Big meteoroids often crack into pieces before eventually burning up. The very largest ones, which may start as metre-sized boulders, give rise to enormously bright meteors. These are sometimes called fireballs and outshine everything in the sky except the Sun and the Moon. In rare cases, even their brightness is challenged, and then we speak of bolides. It is here that we find the predecessors of meteorite falls.

I think that almost every person with good eyesight has been able to see a shooting star, and most of us have seen many. If one takes an interest in the issue and goes to a place with a truly dark sky, the excursion in general does not have to be fruitless at all — it is enough to wait a good while and at least one shooting star should appear. Many people have also seen fireballs, but rather few have had the opportunity to watch a bolide. This reflects the fact that the celestial bodies are more and more numerous, the smaller they are.

There are some occasions when the meteors appear much more often than normally. This phenomenon is called a meteor shower. Obviously, Earth then passes a region in space where the meteoroids are more tightly packed. What can be the reason? The answer was given 150 years ago by Giovanni Schiaparelli (see Section 1.2). He proceeded from the known fact that meteor showers tend to recur periodically on certain days every year and noticed that these days coincided with Earth's passages in the vicinity of the orbits of known comets. His interpretation was that dense streams of meteoroids, originating from the comets, are travelling along these orbits. Earth hence receives a shower of cometary grains at each such passage (see Fig. 3.2).

The most well-known recurrent meteor shower is the Perseids, which shows up yearly close to August 12. The name comes from a perspective effect. All the meteoroids hit Earth from the same direction. The meteors are then seen to radiate away from the point on the sky (the radiant), which denotes this direction. In the case of

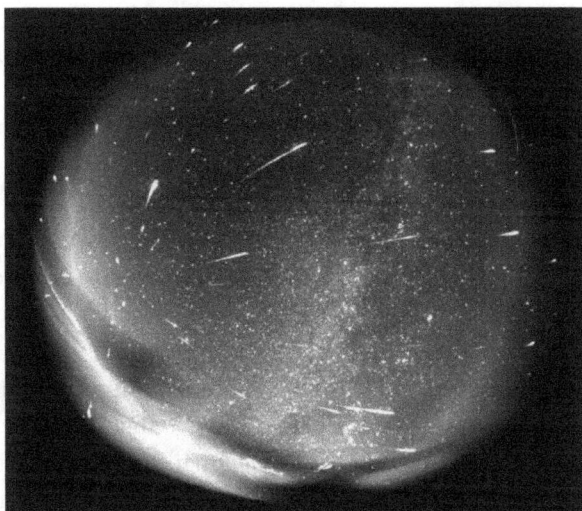

Fig. 3.2. Four hours' exposure of the Leonid meteor shower in 1998, using an all sky fish-eye camera at Modra Observatory in Slovakia. All in all, 156 bolides were reported to be detected on this exposure. Note the diverging pattern formed by these bolides (see the main text). Courtesy: Astronomical and geophysical observatory, Comenius University/Juraj Tóth. License: Creative Commons Attribution-Share Alike 3.0 Unported.

the Perseids, the radiant is situated in the constellation Perseus close to the north celestial pole, which renders it easy to see the meteors from northerly latitudes during nice August evenings. The comet that provides these grains is named Swift–Tuttle and has a period of revolution of around 130 years. Its last approach to the Sun occurred in 1992, and during the years thereabout we hoped to see unusually high numbers of meteors, since the stream of grains should be extra tight in the vicinity of the comet. However, the expected effect more or less stayed away.

We find a different story with the Leonids, which appear around November 17. Their comet is named Tempel–Tuttle and has a revolution period of 33 years. In this case, the comet's approaches to the Sun have often been accompanied by extremely intense meteor showers, when one meteor per second has been sighted. Such events are denoted as meteor storms. The last occasion to watch them was in 1998, and it was marked by a case of scientific progress. The difficulty of predicting meteor storms had seemed embarrassing, and some disappointment around the Perseids in 1992 were fresh in memory. But new computer models for the detailed structure of the meteor stream were produced by David Asher and, for the first time, made it possible to know where the densest passages would exist. Using these, the first really successful predictions were made.

As mentioned earlier, before meteorites fall, a meteoroid usually fragments high in the atmosphere. One can later watch its continued course as a bright fireball or bolide whose motion on the sky is considerably slower than that of usual meteors, since the cosmic velocity is appreciably decelerated. The luminous phenomenon seems to sail across the sky with a "tail of smoke" caused by the abundance of burning fragments (see Fig. 3.3). Eventually, the shock causes an explosion to occur, and a sonic boom can be heard from the ground. Thence, the remainder of the meteoroid may split up further, and a swarm of fragments continues its voyage at such a low speed that these are no longer visible. After this, they finally hit the ground with a velocity that is usually quite modest. They may be cold on the surface, like the air that they had been flying through, but they also exhibit a dark melt crust after the extreme heat experienced at higher levels.

When seeing a bolide, one easily encounters an illusion. Having already seen airplanes in the sky, one knows how fast these seem to move. But although the planes travel much faster than trains or cars, they are practically at rest compared to meteoroids, before the latter have been braked. The bolide thus traverses the sky much faster than

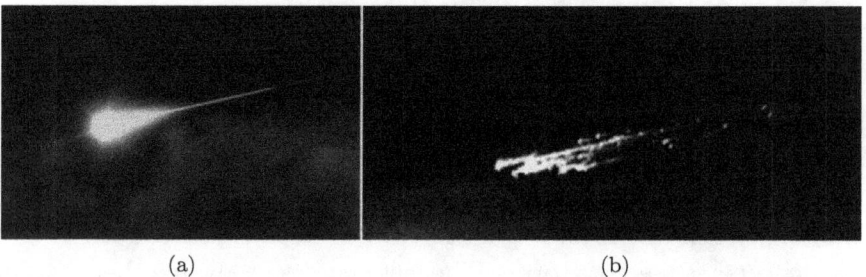

(a)   (b)

Fig. 3.3. Image of a bolide recorded in 2005. Courtesy: APE/Thomas Grau.

airplanes, and one thus gets the impression that it is much closer than it actually is. One may guess that meteorites have fallen very close, though in reality they have fallen hundreds of kilometres away.

One morning in the 1990s, journalists called the Uppsala Observatory, where I was working. Early in the morning, a woman had been riding her bicycle near a local allotment area and witnessed a meteorite fall right in Uppsala. She was sure that the meteorite was lying on one of the allotments. Thus, a whole gang of astronomers went there to search, and I was one of them. When I heard the woman's story, I got a bit sceptical about the search, but we decided to complete it anyway. We found nothing. Later during the day, further observations of the fireball had been reported, and it was judged that a possible meteorite fall would have taken place in the Baltic Sea some 100 km south of Stockholm.

As a matter of fact, not only the very largest meteoroids may reach the ground but also the very smallest. Most particles in space measure just a few microns. Such a small particle feels the impacts by single molecules at very high levels in the atmosphere and is braked without any considerable heating. It may then slowly float down to the ground completely unaffected. If it lands in the ocean, it may be taken care of by sampling the sea floor, and if it lands on the ice sheet of Antarctica, it may reach laboratories for analysis even more easily. In such cases, we speak of micrometeorites. Yet another well-tried method is to collect the particles before they reach the ground with the aid of stratospheric aircraft. We refer to these as "interplanetary dust particles".

3.2. Meteorites: Falls and Finds

There are meteorite collections at most Natural History Museums. They are watched by all kinds of visitors, who hopefully experience the breathtaking feeling of having small pieces of other celestial bodies lying in front of their eyes. But it has not always been evident that such is the case. Leaving aside mystical and mythological notions, until the end of the 18th century, the meteorites were commonly considered as a special, very rare kind of atmospheric precipitation, where rocks took the place of ice or water. The breakthrough came with a book published in Riga in 1794 by the German physicist Ernst Chladni (who died in 1827 in Wrocław, which in the German part of the then divided Poland was called Breslau). In this book, Chladni claimed that at least the so-called iron meteorites have an extraterrestrial origin and that the meteorite falls are connected with bright fireballs.

Like all new, revolutionary ideas, this one too was initially received with scepticism. However, it was supported by chemical analyses, which were soon to be made of several conspicuous meteorite falls. Thus, Chladni was a precursor and a visionary, but if he had not been around, the truth would still certainly have been revealed within a few decades.

The meteorites in the museum collections are of two kinds. Many of them have been seen to fall. The fireballs have been observed, and the fallen fragments were directly retrieved. However, most meteorites have been found on the ground without advance notice. We refer to these two kinds as falls and finds, respectively. In earlier times, the finds were completely random, which means that somebody by accident found a peculiar stone or chunk of metal and brought it to scientists, who deemed it to be a meteorite. In a typical situation, many hundreds or thousands of years had passed since the meteorite fell, so it was often difficult to recognise. For metallic iron meteorites the problem was less acute than for stony meteorites, which might look like any normal stone and thus avoid discovery. For this reason, iron meteorites were over-represented in the collections. However, in recent times a new type of find has appeared, which is much less biassed. This comes from large, organised search efforts at particular places where meteorites have been collected, like in parts of Antarctica, or where nothing has affected them, like in the North African or Arabic desert.

The large number of meteorites that these searches have provided is important, since we have better statistics of some very interesting though rare meteorite types and also found entirely new types. But the falls continue to be of central interest, since they yield the most reliable picture of how abundant the different kinds of meteorites are in space. In addition, they sometimes give us an idea of the orbits that the meteoroids had before colliding with Earth.

As to the respective abundances of iron and stone, the issue is decided. The iron meteorites only contribute about 6% of the meteorites falling each year, and the rest — the vast majority — are stony meteorites, disregarding a small group of 1% where stone and iron are mixed. The stony meteorites, in turn, are of two kinds with different origins. One kind is represented, for instance, by the meteorites from Vesta and Mars, and their origin is broadly volcanic. They represent solidified melt from the crust of a planet or a minor planet with similar properties. The name for these is achondrites and they account for 8% of all meteorites. The remainder are called chondrites, and these constitute the large majority of meteorites. They differ radically from the achondrites by having an origin similar to the sediments from Earth's sea floors without having anything to do with oceans.

At this point, the story becomes really exciting for all who are interested in the origin of the solar system. Indeed, it has been possible to place the sedimentation of the chondrites very accurately in both time and space. I shall not account for the details of this research, but the gist is that this was during the first 1/1,000 of the age of the solar system, i.e. an interval of a few million years, when a rotating, disk-shaped gas cloud surrounded the newborn Sun (see Section 6.1). Solid particles then gathered near the central plane of the disk and agglomerated into the first large bodies (Section 6.2). Part of these pioneers have survived in the form of asteroids and are currently providing us with samples of this original agglomerate in the form of chondrites. We realise that the chondrites can be seen as ancient witnesses of the processes that took place during the very start of the origin of our planetary system. We shall see, though, that many achondrites are ancient, too, and inform us about the same processes in a somewhat different way.

But before we enter into this scientific detective work, I shall present a few of the most interesting meteorite falls that have occurred in my Nordic neighbourhood. On New Year's Day of 1869, meteorites fell near the Hessle estate north of Lake Mälaren in the current community of Enköping. About 700 rocks with a total weight of 20 kg were retrieved within an elongated area measuring about 50 square kilometres. The elongated shape is typical and arises from the fact that different fragments have different sizes and therefore experience different amounts of braking due to the air drag when the fall occurs at an oblique angle to the ground. The largest ones travel the farthest before landing. The Hessle meteorites were chondrites and were easily spotted due to their dark melt crust on the ground, which was white of ice and snow. It has been said that the event made an extra deep impression on the church congregation, who left the morning service impressed by the vicar's sermon just to directly experience the fall of black stones from the sky. In any case, very few of the congregation members ought to have had an idea of what meteorites are.

August 2, 1971, was a fair weather day on the isle of Haverö in the archipelago 25 km southwest of Turku, Finland. A little before 4 pm, people heard thunder-like sounds, but it was so light outside that nobody saw any fireball. Yet, a stony meteorite weighing $1\frac{1}{2}$ kg smashed down through the roof of a storage house belonging to farmer Tor-Erik Andersson about 100 meters from the house where the family lived. This was seen by a nine-years-old boy, who was playing on the yard, and he quickly showed the farmer's family what had happened. The meteorite had landed in a box after having cracked concrete slabs and planks of the roof and perforated the lid of the box.

The rock lying there proved to be a ureilite (see Fig. 3.4). The ureilites are among the most curious meteorites and are also rare. Not many ureilite falls have been observed in the whole world, so the one on Haverö is quite important. This particular ureilite is also among the largest of all and only slightly smaller than the most famous one, which fell in the Russian village Novo Urei in 1886 and is the namesake of the whole group.

A strange anecdote about its fall states that it was partially eaten by the villagers who saw the fall and found it. But the reason that the ureilites are considered curious is that it is difficult to elucidate from where they come. They are counted with the achondrites but do not

Fig. 3.4. A 2.9 cm fragment of the NWA 4231 Meteorite, a ureilite retrieved from the North West African desert. Courtesy: James St. John. License: Creative Commons Attribution 2.0 Generic.

come either from Mars, the Moon, Vesta or any other major asteroid. However, the body they come from has to be relatively large, and it has been held likely that it was once destroyed by a collision. But some light has been shed by another ureilite fall almost 40 years after Haverö, as we shall see in Section 3.5.

Our last stop is on the Danish island of Lolland. In the evening of January 17, 2009, a bright fireball moving westward across the sky was videofilmed by a surveillance camera close to the Swedish town Kristianstad. This was also observed by a meteor camera in Holland and detected by radar from three German stations. Explosions were heard from southern Sealand and eastern Lolland, and plenty of eyewitness reports were registered. It was first thought that possible meteorites would have fallen into the sea, but the German meteorite hunter Thomas Grau had a different idea. Using all the reports and films, it was possible to locate the fall to the region around the small town of Maribo in the midst of Lolland. Grau headed there and, after an intense search, six weeks after the fall, he found a meteorite weighing 30 grams in a cherry garden a few km from Maribo.

The Maribo meteorite too is of particular scientific interest. Indeed, it belongs to a minority among the chondrites that are called carbonaceous and denoted by the letter C (cf. Section 2.2). Different kinds of C meteorites are distinguished by an extra letter, and in the case of Maribo, the notation is CM. The relatively rare CM meteorites are among the most water-rich meteorites and sometimes contain organic substances like amino acids. Maribo is extra interesting, because it was the first for which the observations of the fall were comprehensive enough to determine the orbit of the meteoroid around the Sun. Just a few years later, another CM meteorite fell at a place in California named Sutter's Mill, which is mostly known because the American gold rush started there in 1848. In this case as well, it was possible to determine the meteoroid orbit, and remarkably enough, the orbits of the Maribo and Sutter's Mill meteoroids are very similar to each other.

These orbits fall within a category that is shared by Encke's comet (see Section 4.4) and a number of asteroids. This leads to the question whether the two meteorites may have originated from any of these celestial bodies. If it really were the nucleus of Encke's comet, the comet scientists would find this sensational. In fact, as we have seen, these nuclei appear so enormously fragile that nothing ought to survive a passage through Earth's atmosphere at a high speed. But sensations do sometimes occur, and one should never say never. In addition, the similarity of orbits between comets and asteroids in this category indicates that we may be dealing with special bodies that are untypical of both comets and asteroids — but still exist. Perhaps, a space probe should be sent to one of these so far unvisited objects. If they really are not like the others, they might teach us all the more.

Meteorite finds can also be interesting both due to their large number and specifically in individual cases. The most illustrious is the Martian meteorite ALH84001, which was found in the Antarctic Allan Hills area in 1984. The reason for its fame was statements by scientists about signs of fossil life forms, which were announced in the 1990s. But I leave this aside, since this book is not about planets. Let me instead portray the world's oldest meteorites! These were actually found in Sweden. It must be stressed that meteorites have several different types of ages. In the following discussion, we mean the time that has passed since they fell.

As mentioned previously, meteorite finds in general have ages of about a thousand years. Those that have been found in Antarctica and other dry places can be considerably older. One example is ALH84001, which fell about 13,000 years ago. But no other meteorite comes even close to the Ordovician meteorites. These fell about 470–480 million years ago during the geologic era called the Ordovician. Before they were discovered, a few reports had been written about fossil iron meteorites with a likely age of up to 50 million years and the traces of an iron meteorite in the Ural Mountains that might have fallen more than 200 million years ago. However, a new record was set with a good margin by the Brunflo meteorite.

Ordovician limestone was mined during the 1950s at Rödbrottet near Brunflo in the Swedish province of Jämtland. One day in 1952, the workers were sawing a large block for the typical use in the floor of some railway station. They then noticed a fist-sized, black stone inserted into the block. They had likely seen many a strange thing, but this one was sent to geology professor Per Thorslund in Uppsala for assessment. It was certified that the composition of the stone had a similarity to intrusive, volcanic terrestrial rocks. On the other hand, it did not fit with the expectations of a terrestrial basaltic rock that had happened to be incorporated into the oceanic sediments during the Ordovician. The block was then left without further notice in Uppsala for 25 years, until Thorslund attended to the matter once more. He had been reading about the progress that other geologists had realised by identifying large craters (e.g. the Siljan ring — see Section 5.2) with the results of impacts by asteroids. Thorslund and his Stockholm colleague Frans Wickman performed a close investigation of the stone and found that it showed all signs of being a chondrite that had fallen at the time, 460–470 million years ago, when the Brunflo limestone originated.

Later on, many more ordovician stony meteorites have been found in limestone that is mined in the beautiful region around

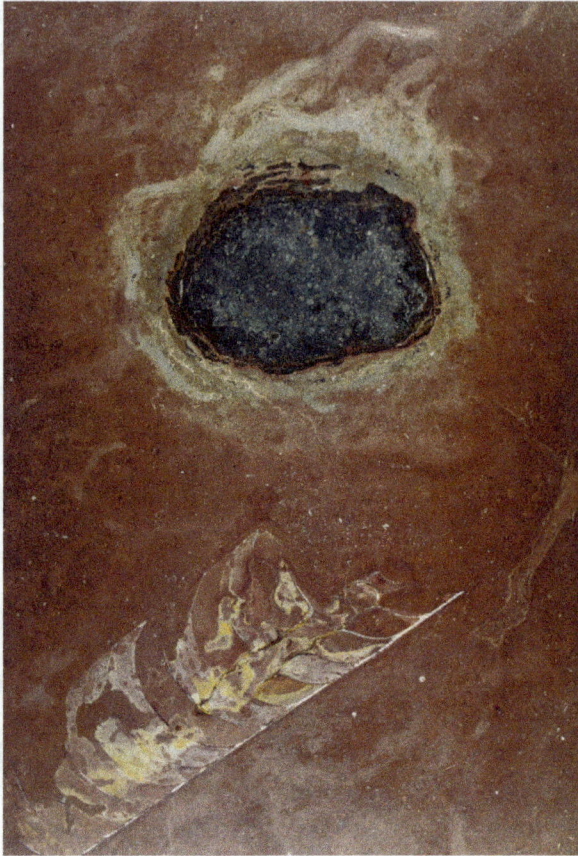

Fig. 3.5. Fossil, ordovician meteorite, exposed next to a nautioloid shell in a limestone from the Österplana quarries in southern Sweden. Courtesy B. Schmitz.

Österplana on the Kinnekulle hill near Lidköping in southern Sweden (see Fig. 3.5). In this connection, a very interesting discovery has been made. The large number of meteorites in this 480 million years old limestone suggests that Earth at that time was struck by a hundred times more meteorites yearly than what we see today. Analyses performed by Birger Schmitz at Lund University and his colleagues have verified that this was indeed the case. What does it mean?

Most likely, the explanation is a large collision, which took place somewhere in the asteroid belt shortly before the Österplana meteorites fell. The meteorite invasion lasted for a few million years in accordance with such a hypothesis. The Brunflo meteorite, however, is more than 10 million years younger and cannot belong to the same shower. It also differs chemically from those of Österplana, which are mutually very similar. The chemistry of the Österplana meteorites fits with a kind of chondrites — the L chondrites, which have another interesting property. They have experienced a more violent shock than other kinds, and this shock can be dated using their content of the radioactive isotope potassium-40 and its daughter isotope argon-40, which is a noble gas. In fact, the shock evacuates the rock of this gas, so that the radioactive potassium clock starts ticking. In this sense, the age of the L chondrites has proven to be between 450 and 480 million years.

One may therefore draw the preliminary conclusion that the parent body of the L chondrites was split by a collision during the Ordovician, which gave rise to the meteorite shower that the Österplana meteorites bear witness to. Such a large collision should also leave behind an asteroid family, and here the large Flora family in the inner part of the asteroid belt seems to fit very well. One member of this family is the asteroid Gaspra, which was imaged by the NASA probe Galileo in 1991 as the first of the asteroids. Gaspra is about 10–15 km in size and is clearly a splinter after a large collision, judging from the pictures.

3.3. The Provenance of Meteorites

We may wonder how long the meteorites had been travelling in space before they reached Earth. The following fact was established long ago. Even though most of them are ancient, being formed in connection with the birth of the solar system, there is no way that they could be vagabonds from that time. In such a case, they would have subsided very long ago by colliding with the Sun or with some planet, or by being ejected from the solar system as a consequence of the gravitational effect of the planets.

There is also a direct way to settle the issue. The surface layer of a meteoroid travelling through space is constantly bombarded by energetic atomic particles in the so-called cosmic radiation. By a careful microscopic study of the meteorites, one can observe the effects of these collisions and estimate the "exposure time" during which the meteorite was subject to the bombardment. Such results show that the stony meteorites have generally been exposed for a few tens of millions of years, while the iron meteorites may have been so for as long as a billion years. In all cases, they have thus spent most of the time incorporated into larger bodies, which have protected them from the cosmic rays.

This confirms what has already been an established consensus: the meteorites are small fragments from large parent bodies. They were only knocked off during the latter part of the history of the solar system and in most cases quite recently. So, which are the parent bodies? Apart from Mars, the Moon and Vesta, which are clearly identified, we have to look among the other asteroids to find the rest. As we saw with the ureilite from Haverö, some parent bodies may have already been destroyed. Others are probably still alive in today's asteroid belt. Large attention has been paid to the issue of which asteroids could be connected to each kind of meteorites, aiming to identify the individual parent bodies or their fragments.

It is evident that the parent bodies were generally large — by way of a guess, at least 100 km. Let us take the iron meteorites as an example. These mainly consist of iron and nickel in the form of alloys. Both iron and nickel are common elements in the Universe and were present in large quantities in the material that the asteroids were formed of. But alloys did not exist from the outset. These must have arisen from a geologic evolution of the respective parent bodies. An extreme heating caused the material to melt, and the melt was separated into different components with different densities. The heavy iron sank to the bottom together with many other metals — there besides nickel. When the heat source disappeared, the asteroid cooled and solidified. In its core, we thus find the origin of the alloys seen in the iron meteorites. The source of the extreme heating may seem mysterious, but I shall return to it in the following section.

Iron and nickel can form two different minerals called kamacite and taenite. Kamacite is very poor in nickel, while taenite may contain up to 50% of it. The first material to solidify at temperatures near the melting point consisted of pure taenite, but as the temperature continued to decrease, plates of kamacite grew inside the taenite. If the nickel content was very low to begin with, only kamacite finally remained. Some iron meteorites are of this kind, while others have more nickel and show a background of taenite with incorporated plates of kamacite. In this case, the thickness of the plates yields information about how quickly the material cooled down. This cooling proves to have been extremely slow — only one degree per million years. Thus, in reality, the parent bodies must have been of considerable size. Too small bodies would have cooled far too quickly (see Fig. 3.6).

The stony meteorites must also come from large parent bodies for the simple reason that, like the iron meteorites, they were formed when the solar system was newborn. Most likely, there then existed bodies of very different sizes, but the small ones did not stand a chance to survive all the collisions that were to follow, and their

Fig. 3.6. Cut, polished and etched face of the Toluca iron meteorite, displaying a Widmanstätten pattern featuring kamacite plates grown inside a taenite matrix. Courtesy: H. Raab. License: Creative Commons Attribution-Share Alike 3.0 Unported.

fragments have also been destroyed since a very long time ago. Only those with at least the size of Lutetia (see Section 2.5) may have contributed to our meteorites, no matter whether they themselves or only their fragments have survived. The fact that the ages of the meteorites can be accurately determined is due to the methods of radioactive dating. One uses radioisotopes and the products emerging as these isotopes decay. The contents of the isotopes are analysed at different spots in the meteorite. From the correlation between the contents of the mother and daughter isotopes, one can deduce the time that has passed since the material was formed.

It is clear that the continuous fragmentation that occurs in the asteroid belt due to the collisions can explain the origin of the meteorites, and this is doubtless the real explanation. But, in addition to identifying the parent bodies, we would also like to understand in detail how the transport to Earth is effectuated. Can the very collisions fling away small fragments at such high speeds that these may travel directly to our planet? No, such speeds cannot be expected. The meteoroids created have orbits that remain in the asteroid belt. But we now come back to the Kirkwood gaps and the associated resonances between the orbits of the asteroids and Jupiter.

Can it be that the parent asteroid is close to such a resonance, and the collision flings particles into the Kirkwood gap? Yes, this can happen, and it surely happens as well. Let us take Vesta as an example. This has a revolution period of about 1/3 that of Jupiter and is thus close to the deep 3:1 gap. It is self-evident that minor crashes into Vesta may toss fragments into this gap, whereupon Jupiter's perturbations lead the fragments into more and more elongated orbits, until these reach Earth. Some of the HED meteorites, which I mentioned in Section 2.5, may have arrived this way. But the impact may also have occurred on some other member of the present Vesta family, which in turn has its origin in the material from the enormous Rheasilvia basin. In any case, it is not a question of the collision that created the basin, because this occurred far too long ago.

The letters H, E and D stand for the three subgroups of Vestan meteorites, i.e. howardites, eucrites and diogenites. The dividing line goes between the eucrites and diogenites, which originate from the upper and lower parts of Vesta's crust, respectively, while the howardites are a mixture of eucrite and diogenite fragments. The observations made by the Dawn spacecraft confirm the idea that the fragments from the large collision represent different layers of Vesta's crust, since the basin reaches almost all the way down to the mantle. The Vesta family thus consists of both eucritic and diogenitic bodies, while the howardites likely come from the rest of Vesta's surface, where small pieces of both materials have accumulated and been deposited like a blanket around the entire body.

The radiation damage of the Vestan meteorites, however, shows that most of them cannot have been thrown directly into the 3:1 resonance. If so, they would have landed on Earth in a shorter time than indicated by the radiation damages seen. The conclusion drawn is that the entry of the meteorites into the resonance was delayed. This delay concerns all meteorites and not only those from Vesta, which served as an example. The journeys from the parent bodies began in peace and quiet with an extended time in the asteroid belt, during which the orbits slowly but surely crept into the resonances by an increase or decrease of the orbital periods.

However, such an effect is not predicted by the planetary perturbations. Hence, another, unknown kind of perturbations seems to be called for. By the way, "unknown" may not be the appropriate word. The basis of the explanation was published already at the start of the 20th century by the civil engineer Jan Jarkowski, born in the present Belarus, who lived and worked in the Russian empire under his internationally known name, Ivan Osipovich Yarkovsky. His work might have been forgotten if the great astronomer Ernst Öpik had not mentioned it long after Yarkovsky's death. What we are dealing with is the so-called Yarkovsky effect. If a meteoroid or asteroid is small enough, and especially if we consider a long enough time, its orbital motion can be affected by its own thermal radiation.

This is reminiscent of the jet force that affects the motion of comet nuclei, although it is here a question of infrared photons instead of leaking gas. The radiation is emitted into the direction facing the hottest part of the nucleus. Due to the rotation of the meteoroid, diurnal and seasonal effects arise, which cause the jet force, in the long run, to point at an angle to the direction opposite to the Sun. This can both accelerate and decelerate the orbital motion, so that the period increases or decreases. About 20 years ago, it was shown that the Yarkovsky effect provides a nearly perfect explanation to the observed cosmic ray exposure times of meteorites.

The hunt for the meteorite parent bodies has, generally speaking, only had limited success, since exact correspondences have not been established with certainty. In spite of this, there are some interesting results to describe. These have to do with the composition of different minerals in different types of chondrites. The ingredients are basically free metals, metal oxides, sulfates, iron and magnesium silicates, hydrated silicates, pure carbon and carbon compounds. The detailed combination of minerals observed in a given meteorite can be interpreted as an indicator of the temperature that prevailed at the time and place of the material's origin.

One then imagines an environment in the infancy of the solar system, where the temperature dropped gradually, and one looks in particular at the minerals that are missing. For some of them, it was too hot so that their condensation was inhibited, while others had naturally existed at higher temperatures but then disappeared through chemical reactions. In reality, it is complicated, since the chondrites are not chemically homogeneous, but some conclusions can still be drawn. We find the highest formation temperatures for enstatite chondrites, so-called E chondrites, where the magnesium silicate enstatite is common and the iron is mostly metallic. Next to these there are a few groups of ordinary chondrites (the most common by a wide margin), which in order of falling temperature are denoted by H, L and LL. The only one of these that has any perceptible water content is the LL group (see Section 2.7).

Based on spectral similarities, the E chondrites are associated with a type of asteroids that is also denoted by the letter E. These E asteroids are mostly concentrated along the inner edge of the asteroid belt. The interpretation of the spectra for ordinary chondrites is much more difficult. The fact that these are divided into three distinct groups may suggest that they ultimately derive from only three parent bodies, but there is no solid evidence of this. On the other hand, the parent bodies or their fragments are no doubt to be found among the S asteroids. These constitute the dominating group in the inner parts of the asteroid belt, but this group is not homogeneous. There are several subgroups, although neither of these has been unequivocally identified as the parent of any of the three types of ordinary chondrites.

As shown by the name, the carbonaceous chondrites often have a somewhat elevated abundance of the element carbon — a property not shared by other types of meteorites — and also in many cases by the appearance of hydrated (i.e. water-bearing) silicates. All of this suggests a lower formation temperature than for the other chondrites. The spectral properties of the carbonaceous chondrites hint uniquely at parent bodies in the outer part of the asteroid belt — primarily the dominant C asteroids but also other variants. There are about half a dozen chemically separate types among the C chondrites. These show significant differences in, for instance, water and metal content, and we certainly have to deal with different parent bodies.

We need to add some extra comments on the parent bodies of iron meteorites. These are evidently destroyed, since we receive small pieces of their iron cores. Many different groups have been identified among the iron meteorites based on their nickel content and the amounts of other, less abundant metals. It is thereby estimated that there were at least a dozen parent bodies from which the iron meteorites originate. Some iron meteorites contain small stony inclusions, where the relative abundances of different oxygen isotopes in the silicate minerals are in agreement with the findings from ordinary chondrites. It thus appears that the same parent body has provided us with both kinds of meteorites, which would imply that its surface layer remained cold and pristine, while the interior became so hot that the iron melted and sank to the bottom.

Whether or not this is the case, there must exist more direct parents of iron meteorites in today's asteroid belt in the form of asteroids with metallic surfaces. It is considered likely that such objects exist among the so-called M asteroids. The large M asteroids Psyche and Kleopatra are among the best candidates. Unfortunately, no space probe has visited such an asteroid, and the proposals of future probes have so far not been successful.

The findings that I have just reported may undeniably seem like important progress — particularly if we consider that the formation temperatures of chondrites inform us on the temperatures that prevailed in the zone of the primordial, rotating disk, where we now see the asteroid belt. But this is perhaps too optimistic. We shall see in Section 6.3 that the young solar system was a very dynamic environment, where bodies may have migrated between different regions. Let me use a metaphor. My wife is of Polish extraction, and as to myself, I am as Swedish as one can be. But looking at the entire Swedish population, there are people of all possible extractions. I think that something similar may be the case in the asteroid belt. However, one difference is that the asteroid belt is quite segregated. The different classes of bodies with different compositions form a systematic sequence from the inner to the outer edge, even though a certain mixing has occurred.

The next major issue is whether there are meteorites originating beyond the asteroid belt. One meteorite has attracted particular attention in this connection. It fell in the morning of January 18, 2000, in northwestern Canada. From an area in Yukon and British Columbia, a fireball was seen to explode at a high altitude. More than 500 meteorites with a total weight of more than 10 kg were retrieved from the snow-covered ice on Tagish Lake. Together, these are referred to as the Tagish Lake meteorite. It is estimated that more than one tonne of material reached the ground, but that 97% of the meteoroid was vaporised and dispersed in the atmosphere. This indicates that the meteoroid was very fragile, and the retrieved material is highly porous. It reminds of the carbon briquettes used in grills and boilers.

The Tagish Lake meteorite is classified as a carbonaceous chondrite but is not like any other. Chemical analysis has shown that its closest relatives are the most primitive carbonaceous chondrites (types CI and CM), whose elemental composition is most like that of the Sun. It is rich in organic substances like amino acids, and the carbon often appears as nanodiamonds.

It certainly has its own parent body, which differs from all the others. A possible clue in the search for this is the fact that the spectrum of Tagish Lake is similar to spectra from a special group of asteroids in the outer part of the belt — the so-called D asteroids. It is as yet a question of speculations, but Tagish Lake may very well be a fragment of a D asteroid. If so, we have one prominent candidate: the nearly 100 km-sized Irmintraud. It has the advantage of being very close to the 5:2 resonance with its deep Kirkwood gap. Hence, it is easy for pieces of Irmintraud to reach Earth.

But the real home of the D asteroids in the solar system appears to be situated some distance beyond the asteroid belt. They dominate among the asteroids in this region, where we find the Hilda group in 3:2 resonance with Jupiter and the trojans, which were presented in Section 2.2. No meteorites can reach us from these groups owing to Jupiter's influence on the orbits. Bodies like Irmintraud, which may once have migrated into the asteroid belt, could thus possibly give us our only chance to analyse this interesting material.

A major issue is, of course, if meteorites can come from comets. I touched upon it in connection with the Maribo meteorite, but I left it too quickly. As a matter of fact, both D and C asteroids may have a relationship to comets, independent of hints coming from single meteorites, so the issue is burning. Surely, we have seen that comets are extremely fragile, and as I mentioned, it is difficult to believe in their connection to meteorites. However, in the light of the Tagish Lake case, one may still glimpse a chance. Almost everything disappeared in the atmosphere but not 100% — a few percent actually did reach the ground.

One circumstance is often considered to speak against comets as parents of meteorites. During centuries' worth of observations of meteor showers originating in comets, lots of bolides have been seen in the sky without a single meteorite being found. For sure, this is true and requires an explanation, but it might be as simple as the shower meteoroids hitting the atmosphere with too high speeds. The comets in question, of which Halley's comet is one example, have orbits around the Sun that are strongly inclined to Earth's orbit or even oriented in the opposite direction. Therefore, the velocities have to be very large and the meteors become bright. If one would take such a meteoroid and throw it into the atmosphere at a low velocity, typical of the meteorite falls, perhaps meteorites might survive.

Among the most famous meteorites, there is actually one that might come from a comet. The Orgueil meteorite fell on May 14, 1864, in southern France, and is the chemical archetype of the CI meteorites. It provides the best example for the similarity between the elemental abundances of chondrites and those of the Sun, and it is rich in water, bound up in hydrated silicates. More than 10 years ago, a team of scientists headed by Mathieu Gounelle embarked upon a determination of the meteorite's orbit with the aid of the old reports of fireball observations. They found that this orbit likely reminds of the type of cometary orbits that characterise most targets of space missions: Wild 2, Tempel 1, Hartley 2 and Churyumov–Gerasimenko.

If Gounelle and his colleagues were right, even though this is far from being agreed upon, the Orgueil meteorite becomes more interesting than anyone had imagined. It would represent 14 kg of the surface crust of such a comet. Pity, though, that we have no chance of establishing which comet it is. But I cannot help noting that what we have seen on close-up images of comet nuclei — in particular, the Rosetta comet — in no way denies such a possibility.

3.4. The Allende Meteorite

The first lunar landing took place more than 50 years ago. It was on July 20, 1969 that Neil Armstrong took the famous step from the Lunar Module Eagle onto the surface of the Moon during the Apollo 11 mission. Thereby, part of the "space race" had a winner, but what does this victory mean today? For me, science is the most important, and it is easy to realise that the retrieval of, in total, 382 kg of lunar rocks and gravel from the six Apollo landings has been of revolutionising importance for our understanding of the solar system's history.

This progress is certainly worthy of admiration, but one should not forget a completely unexpected side effect, which arose due to a meteorite fall. Awaiting the arrival of the lunar rocks, new instruments with sensitive measurement devices were constructed for chemical analysis and the determination of isotopic abundances. New methods were developed for the use of these instruments, and lots of young scientists were educated, aiming for this major effort. It was of course largely at the NASA institutes in the USA that all was ready for receiving the lunar material. Then, on the night of February 8, 1969, a giant meteorite fell like a gift from above.

It was in the province of Chihuahua in northern Mexico that the fireball was witnessed. The strewn field of the meteorites was quickly identified. It was situated around the village Pueblito de Allende, from which the name Allende meteorite derives (see Fig. 3.7). If this had been a common meteorite fall, it would not have been very important. But the Allende meteorite is by far the largest carbonaceous chondrite that has ever been seen to fall. The retrieved fragments altogether weigh more than two tonnes, which is more than five times the total weight of the Apollo rocks. That the fall occurred in a desert was also important in order for the fragments to be conserved and easily identified. Finally, the location of the desert close to the NASA laboratories was optimal. The Gobi or Kalahari deserts or any other one would not have been as convenient.

The Allende meteorite is a CV3 chondrite and, as such, contains almost no traces of water. After sawing off a fragment and polishing its cut surface, one gets a very good picture of the structure of the chondrites and the typical components that are, in principle, present in all of them. Most typical among the components are the chondrules — round, glassy grains with a typical size of a few millimetres. The word comes from the Greek *chondros*, which means seed, and from this we derive the name "chondrites" for the meteorites in question. I earlier referred to chondrites as a kind of

Fig. 3.7. Allende meteorite specimen, featuring CAIs as the light coloured inclusions. Courtesy: Natural History Museum of Denmark. Courtesy M. Bizzarro/Z. Fihl.

sedimentary rocks. Chondrules are immersed as guest particles into this sediment. Before the sedimentation took place, as far as we can see, they were stony grains that were suddenly heated so that they melted and quickly solidified again. The reason for this is not established and is vigorously debated among scientists.

Another frequently occurring component consists of light, irregularly shaped grains, which can reach up to several centimetres in size. These have an extreme chemical composition, since they are formed exclusively of high temperature minerals. By this we mean minerals that can exist in solid form, withstanding evaporation, even at very high temperatures. The aluminium oxide corundum (Al_2O_3) and the calcium titanate perovskite ($CaTiO_3$) are typical examples. The common denomination for these grains is CAI, i.e. *Calcium-Aluminum-rich Inclusions*. How can such grains have arisen in the newborn solar system? Clearly, this happened at a place with a very high temperature, and this indicates the region closest to the Sun or a very early stage in history — perhaps both.

The timing is in any case well-defined. By radioactive dating, one has measured the age of the CAI grains at 4.567 billion years, which exceeds the ages of all other solar system materials. Thus, the Allende meteorite with its enormous mass is of great importance. Unfortunately, though, we have no idea about where the parent body resides, except that it ought to be an asteroid in the outer parts of the belt. The origin of CAI is considered to mark the birth of the solar system, and other events are usually dated with respect to this zero point. But as if this would not be enough, yet another discovery of enormous importance has been made with the aid of the CAI grains in the Allende meteorite.

An American research team found that a magnesium isotope with 26 mass units (not the normal one) is enriched at sites with large aluminium content. The correlation between the degree of enrichment and the aluminium content is so good that there ought to be an explanation. In nature, aluminium exists at nearly 100% as one single isotope with 27 mass units. But in the atmosphere there are traces of the lighter isotope Al-26, which is formed when argon atoms are split by cosmic rays. This isotope is unstable with a half-life of 720,000 years and decays into the stable magnesium isotope Mg-26. The explanation for the enrichment of Mg-26 in the Allende meteorite is that its aluminium originally had a significant fraction of Al-26, which became replaced by Mg-26 within a few million years.

The big issue is, from where did this radioactive aluminium come? We can immediately exclude a massive destruction of argon in the early solar system. Instead, the aluminium must have been formed somewhere in the vicinity of the solar system and then found its way into the rotating gas disk around the growing Sun. As a consequence, it also entered into the first solid grains that we identify with the CAI of chondrites. But how did this occur? Al-26 is formed by nuclear reactions taking place in the inner parts of massive stars — considerably more massive than the Sun — as these approach the end of their short lives. It then escapes into space, either through a wind that blows from the star when it is still alive, or together with the other remainders of the star, when this explodes as a supernova. The conclusion is that such a star was situated close by, as the Sun was born.

Now, shall we regard this as an incredible coincidence or something natural? Scientists are reluctant to believe in coincidences that are incredible, so there is a drive to find a logical connection. Different ideas have been proposed. Today, the dominating view holds that the Sun was born out of a minor part of a giant cloud of gas and dust, which gave rise to plenty of stars. Such clouds are observed in the Milky Way, e.g. in connection with the Orion nebula, and these are commonly called *Giant Molecular Clouds* (GMC). Among these lots of stars, which were born in the course of a few million years, there should be some that were born early and with a large mass. These may have died in time to pollute the gas that later stars were formed of by residuals like Al-26.

When Al-26 decays into magnesium, energy is produced. High-speed particles are emitted from each atom that decays and share their kinetic energy with other atoms and molecules. This heats the material. As long as a centimetre-sized CAI grain is moving freely in space, the temperature of the grain can only rise insignificantly, because the extra heat is easily radiated away. But, as mentioned, there came a time when sedimentation occurred and the grains were captured into the depths of large bodies, which became the parents of smaller asteroids and meteorites. If this happened early enough, so that large enough quantities of Al-26 still remained, these parent bodies were heated to the point of melting. This is believed to have caused the origin of the iron meteorites and the volcanism that occurred on Vesta and other large bodies.

We see that the Allende meteorite has had a large influence on the research around the origin of our solar system. I wonder where we would be today if it had fallen not in 1969 but more than 50 years later!

3.5. Almahata Sitta

Something very spectacular happened in October 2008. As usual, a number of astronomers in the USA and around the world were busy surveying the sky in search of certain small asteroids that pose a threat of colliding with Earth. This project will be described in some detail in Chapter 5. One of the observing sites was Mt Lemmon in the Catalina Mountains north of Tucson, Arizona, where the Catalina Sky Survey was carried out. During the night preceding October 6, Richard Kowalski was working there. He discovered an interesting object that was directly reported to the Minor Planet Center in Cambridge, Massachusetts. This constantly operating centre is devoted to verifying and forwarding such observations and predictions based on these to other places around the world for follow-up.

Kowalski's object received the designation 2008 TC$_3$ and was promptly observed from other places. The first, automatic orbit computations showed that it was very close and, moreover, heading straight toward Earth. This was the first time in history that an asteroid was discovered almost immediately before colliding with Earth. It was quickly confirmed that the impact would take place about 19 hours after the discovery. The impact site was localised to northern Sudan, where the time would be early morning on October 7.

When this information was distributed, many astronomers were at work with different telescopes so that they were able to carry out important, detailed studies of the asteroid. Its size was estimated at only a few metres, but it was so close that the brightness would still be sufficient. Some astronomers were unfortunately clouded out, and for others the asteroid was too close to the horizon to point the telescopes. But some British astronomers on the Canary Island La Palma were both lucky with the weather and reactive enough to secure the spectrum of the asteroid. This proved to be both unusual and interesting. The closest correspondence has been found among the dark B asteroids, whose chemical nature was practically unknown.

Dawn broke on October 7 over northern Africa, and the impact of the asteroid in the Nubian Desert was witnessed in different ways. The bolide and the smoke trail left behind were seen from far away by an airplane crew. The weather satellite Meteosat also recorded the event with its infrared camera. The lightning of the explosion was caught by a web camera more than 700 km to the north at the

Fig. 3.8. The search for Almahata Sitta meteorites. This picture shows Petrus Jenniskens spotting a fragment on February 28, 2009. Courtesy: NASA/SETI/P. Jenniskens.

Egyptian seaside resort El-Gouna. Acoustic receivers as far away as Kenya detected the sound.

The next important task was to try and find the meteorites. This work was organised by the US-based Dutch astronomer Petrus Jenniskens together with Muawia Shaddad from the University of Khartoum. Both of them set out on an expedition into the desert, accompanied by a large number of Shaddad's students (see Fig. 3.8). They promptly found many meteorite fragments at a place called Almahata Sitta, which in Arabic means "Station Six". Thereby, a new record was set. Never before had meteorites been found after a fall where the meteoroid had been observed as an asteroid before the impact. But we may now note that the fall in Almahata Sitta was the first but not the last time that this occurred.

On New Year's night of 2014, Kowalski made a new discovery at the same place and with the same telescope as in October 2008. This time, however, much fewer observations were made, and the orbit of the asteroid (2014 AA) and the trajectory of the asteroid could not be as well-determined. The size of the asteroid was similar to that of the earlier object. An imminent impact seemed inevitable, but it was impossible to localise the site accurately enough. Estimates that seemed reasonable stretched from Central America to Yemen. Most likely, the asteroid fell into the Atlantic Ocean.

On June 2, 2018, it was time again. Once more, the same astronomer and the same telescope were responsible for the discovery. The asteroid was again of the same size and was designated as 2018 LA. The Mt Lemmon observations only extended over 85 minutes but sufficed to conclude that an imminent collision was very likely. Some further observations were made from Hawaii, when night came there, and these confirmed that an impact would take place in southern Africa. A report of a very bright bolide came from southern Botswana in the afternoon of the same day. A search for meteorite fragments was made in Namibia and three weeks after the impact resulted in what is likely a few pieces of 2018 LA. At the time of writing, I have no further information about these pieces.

We return to the Almahata Sitta meteorite. In total, over 600 fragments were found with an aggregate weight of more than 10 kg. Their composition is enigmatic to say the least. The large majority are ureilites, i.e. relatives of the Haverö meteorite. But the remainder are totally different, being made up of E chondrites and ordinary chondrites of types H and L. Already the fact that chondrites may coexist with achondrites in the same, tiny celestial body (2008 TC_3 measured about four meters) is very surprising, and here, in addition, we deal with several different kinds of chondrites.

As to the solution of this puzzle, one interesting suggestion has been put forward. The parent body of the meteoroid must be an asteroid whose surface layer is a mixture of small pieces made of different materials. Probably, this may emerge if two asteroids with different compositions collide at a low speed and small splinters of both are mixed together. Such a collision is most likely to occur between members of the same family, and hence the family in question has to be chemically heterogeneous. Actually, the inner part of the asteroid belt happens to host such a family, which in addition fits well to the orbit of 2008 TC$_3$. This is called the Nysa-Polana family and has two large members: Nysa that is an E asteroid, and Polana that is a B asteroid. Furthermore, the family also involves S asteroids, so that parents of all the Almahata Sitta fragments are present there.

It is conceivable that 2008 TC$_3$ comes from the surface of one of the smaller members of the Nysa-Polana family. This may have inherited an inhomogeneous composition and, in addition, swept up small fragments from collisions between other family members or collisions involving the locally dominating S-type asteroids. This picture may not necessarily represent the truth, but in any case, it serves to indicate that the asteroid belt may have been the scene of unexpected events, which created asteroids with unexpected properties.

Allow me a final speculation. It is possible that all E chondrites are connected with Nysa and all the ureilites come from Polana or its predecessor. In that case, I wonder if the woods around Haverö may host more fragments from the same fall and if these may exhibit the same diversity of types that was found at Almahata Sitta. However, to my knowledge, no one has made any serious search.

Chapter 4

The Evolution of Comets:
Capture, Fake Death and Outbursts

The most characteristic property of comets is their transience. This is manifested during each revolution around the Sun, when the comet brightens up and then fades away as the distance to the Sun varies. Then, the comet only reacts to changes in its environmental conditions — in particular, the intensity of sunlight — but as a consequence, it also changes at depth. Thus, a comet can never be expected to remain the same from one occasion to another if thousands or millions of years pass in between. Very often, much shorter intervals are sufficient. This means that it is important to understand their evolution if we are to be able to relate the comets observed today to the comets that existed when the solar system was young.

The changes concern both the orbits and the nuclei of the comets. With the aid of computers, one can trace the motions of comets forward and backward in time. When this is done for a large sample of comets, one may obtain a reliable picture of the probability of different orbital evolutions. Exploring in this way the future of all the comets that have been observed during the last century, one finds that almost all will likely leave the solar system within a million years. How shall we interpret this, considering that the solar system is billions of years old and not only millions? In principle, one can think that we live in a privileged time — there have never been any comets in such orbits like we observe today, and soon they will all be gone. However, this idea begs the question: Why is this happening right now? Why are we privileged? The only answer would be that this is a curious coincidence.

In science, we have an aversion to such coincidences, and we thus work with the hypothesis that comets like those of today have always existed and have always been transferred between different regions: a source region where they were formed, a distant reservoir where they have been dwelling most of the time since then, the orbits of the currently observed comets, and finally, the immense interstellar space of the Galaxy, where they likely will eventually be deposited. Investigating the transport routes that the comets follow and the underlying mechanisms is an interesting and important part of comet research.

Another problem concerns the limits to survival of comet nuclei against the processes that may destroy them or put an end to their characteristic activity, which distinguishes them from asteroids. It is possible that the nucleus is simply consumed by erosion owing to the loss of gas and dust that occurs each time it approaches the Sun. Halley's comet lost about one part in a thousand of its mass at its passage in 1986, and more or less the same was the case with the Rosetta comet Churyumov–Gerasimenko during its recent, scrutinised return. If this would continue at a constant rate, their remaining lifetimes would be fairly short. This situation may be changed if the orbits change so that the comets no more come as close to the Sun, or if the surfaces of the nuclei become clogged so that the gas flow is stopped. But in both cases, we lose the bright comets that we now can see.

We seek to understand what is required for the loss of comets to be balanced, in the long run, by an infeed of new comets from their distant reservoirs. How do the nuclei evolve? What distinguishes the rocky surface of Churyumov–Gerasimenko from that of a fresh, pristine comet? How large is the rate of supply from the reservoirs into the group of observable comets that is needed to explain the number of comets that we see? What does this tell us about the size and structure of the reservoirs?

4.1. The Oort Cloud

In Chapter 1, I wrote about two very bright comets: West in the 1970s and Hale–Bopp in the 1990s. Both have orbits that reach far beyond the limits of the planetary system. It is possible to compute in detail how they have moved on the way toward the Sun and also how they will move on the way out. In this way, we obtain the orbits that the comets had and will have in relation to the centre of mass of the solar system. We thus obtain values for their revolution periods. The original revolution period of comet Hale–Bopp was about 4,270 years, and thus, its preceding passage may possibly have been seen by the ancient Egyptians. The next revolution will only take 2,400 years, and hopefully, some people will be able to see it on the next occasion, too.

The times are longer for comet West. Its former revolution took 16,000 years, and the new one will take as much as 6.4 million years. We see that the period of revolution can change dramatically as the comets round the Sun, and the cases of West and Hale–Bopp are fairly typical. In general, the perturbations can be both larger and smaller than for these, and the reason is primarily Jupiter's gravity. The period of revolution reflects the energy that binds the material of the comet to the solar system, i.e. the energy that has to be supplied for the comet to leave the solar system and escape into the Galaxy. Evidently, comet West is on the verge of being ejected. It may possibly return, but even this is hard to say with confidence. That it has a period of revolution of 6.4 million years hence has to be taken with a large grain of salt.

Let us regress to the time of my birth at the end of the 1940s. In those days, the astronomers had no electronic computers to assist them in their calculations. Their tools were utterly primitive, and complicated calculations required a lot of time and effort. Yet, there was a long-standing interest in the type of calculations that I just mentioned for comets West and Hale–Bopp. One simply wanted to make sure if the comets belong to the solar system or are captured guests from the Galaxy. When the 20th century was still young, astronomers Elis Strömgren and Gaston Fayet published such results. It was thus shown that the original orbits of all the investigated comets had elliptic shapes and that the comets were thus bound to the solar system.

The number of investigated comets was very small, but the results interested the famous astronomer Jan Hendrik Oort at the Leiden Observatory in Holland at the end of the 1940s (see Fig. 4.1). His young colleague, Adrianus Van Woerkom, had studied the effects of the perturbations caused by the planets on the binding energies of comets. Oort wanted to see how well Van Woerkom's results agreed with the orbits of observed comets and therefore collected the results from the old investigations. These concerned 19 comets. When he saw how the binding energies of these 19 comets were distributed before the comets reached the planetary system, he must have been greatly surprised.

The most natural idea — or prejudice — would be that the comets had been formed in the inner parts of the solar system and then had their orbits perturbed by the planets so that the energy grew and the comets gradually started to leave the solar system. According to Van Woerkom, if this were the case, the energies of different comets would be uniformly distributed up to the value zero,

Fig. 4.1. Jan H. Oort (1900–1992), Dutch, world-leading astronomer. Credit: J. van Bilsen. License: Creative Commons CC0 1.0 Universal Public Domain Dedication.

where escape into the Galaxy would occur. But the distribution that Oort saw was in stark contrast to this expectation. About 14 of the 19 comets were found in a very narrow interval next to the value zero, corresponding to orbital periods longer than 1 million years. Only five were scattered more or less evenly over all the other values.

How could this be explained? There is absolutely no chance for the planetary perturbations to make the comets pile up inside such a narrow interval. The only option is that there is a source from where the comets arrive in just the kind of orbits that the 14 comets had. Since such orbits extend to distances of at least 20,000 astronomical units, i.e. 20,000 times the Sun–Earth distance, the source should be situated very far away. As the orbits are randomly oriented, the source also must have a roundish shape. This was the suggestion that Oort made in a famous publication from 1950, and the distant comet source has since been called the *Oort cloud* (see Fig. 4.2).

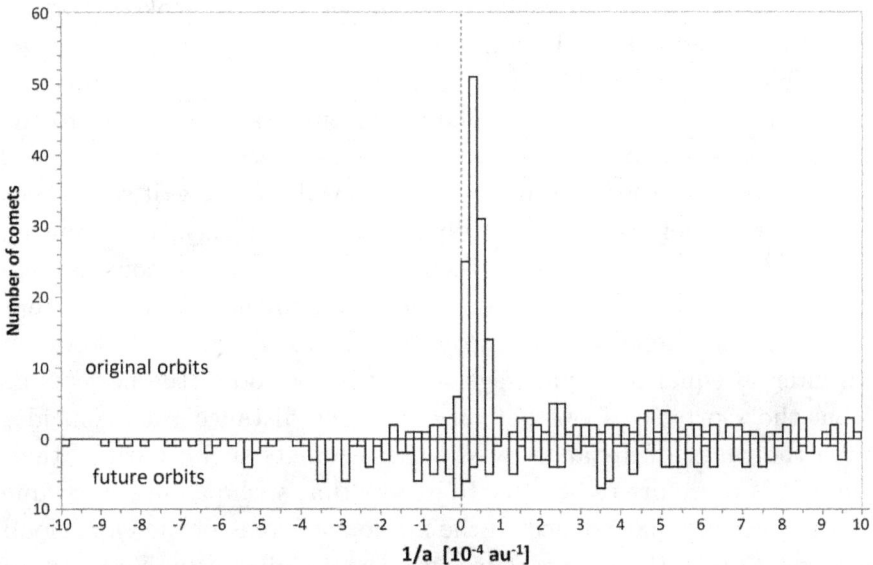

Fig. 4.2. Histogram plot showing the distributions of total energy for one sample of high quality comet orbits. The upper part refers to the energies that the comets had before entering into the planetary system, and the lower part similarly refers to the orbits of the same comets after leaving the planetary system. Courtesy T. Wiśniowski.

Nowadays, Oort's 19 comets have been supplemented by hundreds of other comets, but the impression that he had still stands. In a histogram of the cometary binding energies, one sees Oort's narrow peak rise like a skyscraper above the surrounding small house settlement. If one also shows how the future energies of the same comets are distributed, one realises that the skyscraper falls because of the perturbations imposed by the planets (especially Jupiter). About half the newcomers from the Oort cloud are thrown away into the Galaxy, and the other half will return in a distant future with an energy distribution reminding of the small house settlement in the diagram of original energies. In our somewhat sloppy, scientific jargon, we usually refer to new and old comets. Those in the narrow peak are called new, and the other "revenants" are called old. In general terms, this is correct, but not in individual cases. As we have seen, comet West may perhaps return in 6 million years. In such a case, it will look like a new comet even though it is actually a revenant. There may be such cases in the narrow peak of current comets, but these must be a minority.

What is it that makes comets in the Oort cloud leave the safe life far from the Sun and the planets to embark upon a trip to the innermost region of the solar system? Oort had a very good idea. Along the Sun's route around the centre of the Galaxy, it sometimes encounters other stars. These fly past at speeds typically amounting to 40−50 km/s (kilometres per second, not per hour!) at considerably closer range than the other stars. The longer the interval of time that you consider, the closer the encounters are expected. If we settle on an interval equal to the average revolution period of the newcomers from the Oort cloud, we get an encounter distance that coincides with the range of distances where these comets spend most of their time. It is therefore reasonable to expect that such passing stars, due to their gravity, have deflected the motions of some of the Oort cloud comets, so that these have fallen into the vicinity of the Sun and we have been able to see them.

Oort also argued that all the stellar passages that have occurred over the age of the solar system ought to have had a joint influence on all the comets in the cloud. The comet orbits should thus have been randomised into all possible shapes from circles to the extremely elongated ellipses of the new comets. Based on this, he could estimate the number of comets at about a 100 billion. This estimate would be relevant to explain the number of new comets observed on average each year. If one took a reasonable estimate of the mass of a typical comet according to what was believed at that time, the mass of the entire cloud might be comparable to the mass of Earth — perhaps even larger. This means that the Oort cloud is thousands of times more massive than the asteroid belt. The aggregate mass of all the comets is so large that they may have affected the planets in a much more significant way than the asteroids.

By coincidence, both Whipple's work on the solid comet nucleus and Oort's work on the origin of comets were published in the same year, 1950. The two remain cornerstones of all recent research about the comets in spite of the fact that both Whipple and Oort had to endure bitter criticism from those who would not be convinced. As to the nucleus, irrefutable evidence would later come in the form of close-up pictures, but there is no such evidence for the Oort cloud. There are still doubters, but the arguments are so watertight that they carry as much weight as pictures.

Both Oort and Whipple grew very old, and I had the honour of meeting both of them. I took part in the celebration of Whipple's 90th anniversary in Cambridge, Massachusetts, where he lived and worked. He still rode his bike to the observatory and lived almost 10 years more. When I gave a seminar talk at the Leiden Observatory around 1990, I had Jan Oort in my audience, and he was then about 90 years old.

As seen in Chapter 1, today we have acquired an extreme amount of knowledge about comet nuclei. Something of the sort can be said about the Oort cloud, too. In the 1980s, it was established that there is a different, more efficient mechanism of reshaping the Oort cloud comet orbits and feeding new comets into the Sun's vicinity than Oort's single stellar encounters. This is the collective, gravitational effect of all the stars in the Galaxy. We have known for a long time that the Milky Way Galaxy has the shape of a rather thick disk and that the Sun belongs to this disk. The comets in the Oort cloud feel the slight difference between the gravity of the disk at the place of the Sun and the place where they are situated. This is referred to as a Galactic tide, and it continuously changes the shapes and orientations of the comet orbits.

After the discovery of the Galactic tidal effect, it became the king of theories about the Oort cloud, and Oort's single stars were more or less considered obsolete. But they were to come back — in part through my own research. In fact, the tidal effect works splendidly for the part of the comets whose orbits have certain, specific properties, but not at all for many others. If one estimates that the Oort cloud was formed with all kinds of comet orbits, which is usually done, all is in order — at least to begin with. But in the course of time during billions of years, the special comets may end up. We see what happens today. Almost all the new comets are lost. Some are thrown into the Galaxy, never to return. Others are captured into stronger bound orbits (I shall discuss these in the following section). They are in any case lost from the Oort cloud.

Since the Oort cloud was certainly formed at least 4 billion years ago (see Section 6.4), the existence of the new comets requires that there is something that can operate on the Oort cloud so that comets are transferred into the special orbits that are sensitive to the Galactic tide, thereby replacing those that were lost. This is indeed the perturbations by single, passing stars. This explanation actually implies a collaboration or synergy between single stars and the Galactic disk. The phenomenon may be compared to a soccer team. We identify the appearances of new comets with the goals scored in a soccer game. This scoring is often effectuated by the famous stars like, e.g. Messi or Ronaldo, and the tidal effect plays this role in the case of the Oort cloud, but in the long run no goals would be scored without the efforts by the whole team.

There is another development from the 1980s, originally due to astrophysicist Jack Hills, which may be even more important. The fact that the new comets from the Oort cloud have orbital periods of millions of years does not tell us much about the structure of the cloud. If there would be an inner core, where the comets have much smaller orbits and shorter periods, we would not be able to see this, because such comets are insensitive to the gravitational influence of normal, passing stars and to the Galactic tidal effect as well. We would likely have to wait for a 100 million years, until any star passes through the inner core. If this happens, a very strong shower of new comets would appear. However, there is neither any evidence that such showers have appeared, nor of the opposite.

One argument for the existence of an inner core came from the realisation that comets in the region that the new comets come from live a risky life. The Giant Molecular Clouds (see Section 3.4) had been mapped by radio astronomers as to their number and locations in the Galaxy. It was then realised that the Sun must have passed close to such giants a number of times, and on these occasions the comets could have been affected strongly enough to become unbound and leave the solar system. If the past history has really been as bad, there would have to be a stable reservoir to fetch new comets from, and the inner core might fulfil this function. Yet another argument came from computer simulations of the origin of the Oort cloud, which started to be carried out in the 1980s. I shall come to this subject in Chapter 6, but we may note that already the first results showed a likely structure of the cloud where the inner core is a dominating component.

4.2. The Road to Halley

What happens to the revenants? We have seen that Jupiter's gravity perturbs the orbits of all the new comets from the Oort cloud. These orbits have several important properties like, for instance, their inclination to the plane of planetary orbits or the smallest distance from the Sun, but Jupiter's largest influence is on the binding energy. This may increase or decrease, and both outcomes are about equally likely in general. Half of the new comets are directly ejected into the Galaxy, since their energies already were in dangerous proximity to the critical limit. The others constitute the revenants, which are to return in more strongly bound orbits. When this happens, they are again perturbed by Jupiter. If they are not ejected that time either, they will come back a third time, and the same story will be repeated.

Thus, the binding energy is subject to a series of jumps, and this series has an important property. There is no correlation between the different, consecutive jumps as to their size or direction. The revenants do not have any tendency to evolve constantly toward orbits that are bound more and more strongly. Only coincidence decides where they will eventually go. With a picture from the animal kingdom, we may imagine a livestock transport, which delivers a herd of kangaroos to a place at the edge of a cliff. They of course start jumping but can only jump to the left or right. On the left there is the precipice, and this is the end for every poor kangaroo that enters it. But the kangaroos are blindfolded and cannot see where they jump.

The random walk can start. Directly after the first jump, half are gone. At the next jump, some more disappear, and so it continues. As time passes, fewer and fewer kangaroos remain. Among the lucky survivors we principally find those who were fortunate enough to take a really long jump to the right already the first time. The very few who took several such jumps before the first one to the left may be far from the precipice and thus fairly safe.

We return to the comets. In fact, Oort made yet another important observation. The contrast between the skyscraper and the small house settlement is too large. If the revenants would have the same properties and shine as brightly as the newcomers do, we ought to see much more of them. Identifying the reason for the lack of revenants is a classical problem in comet research. Different suggestions have been put forward, but all of them have their weaknesses. My American colleague Paul Weissman claimed long ago that the comets split and hence disappear, but I think that if this is true, we should see considerably more of those splits. Other theories imply that the newcomers are special by coming for the first time from billions of years in cold storage in the Oort cloud.

One may thus expect that the surface layers of their nuclei have suffered radiation damage due to energetic particles coming from the Galaxy that have bombarded them with so-called cosmic radiation. The effect would be that the surface is covered by a crust of carbon-rich polymers that prevents evaporation of the ice.

On the other hand, this crust might be cracked by interior gas pressure due to very volatile species like carbon monoxide. Lots of dust grains would be produced by the cracked crust and render the comet extra bright due to the same scattering of sunlight as we find in all comets.

I tend to think that everything that may make the comets lose more and more of their brightness in the course of time may also help solve the problem with the lack of revenants. The only caveat is that, as long as the nuclei survive, these might still be observable with the large survey telescopes that have been used in recent times. It has been pointed out that the dearth of such detections may mean total disappearance rather than fading of the comets, but I am sceptical.

In any case, if the new comets do flare up during their first close approach to the Sun, this may help explain a circumstance that has been somewhat awkward for astronomers. On some occasions, new comets were discovered at large distances from the Sun years before the closest passage. These have been extremely bright in relation to the distance. One such comet was discovered on March 7, 1973, by the Czech–German astronomer Luboš Kohoutek. When its brightness soon after discovery was extrapolated to the time around the forthcoming approach to the Sun, one saw a chance of a magnificent spectacle in the sky, and newspapers wrote about "the comet of the century". Some thought that the world's demise was imminent.

When the time of the spectacle arrived around the turn of the year 1973–1974, Kohoutek's comet shone with respectable brightness, but it was far from living up to the high expectations. People saw it as a big disappointment, and the astronomers had made a fiasco. The explanation may be that the comet had flared up at the time of its discovery, perhaps for the reason outlined earlier, and thus showed an anomalous behaviour but thereafter returned to the normal cometary behaviour. At that time, the physical mechanisms that might explain such anomalous behaviour were not as well understood as now, and there was hardly any previous experience to compare with. Thus, the risk of Kohoutek-like "fiascos" was larger.

Notwithstanding the fact that old revenants are relatively difficult to discover and sometimes may have split up and disappeared, we can also note that among the observed revenants, some appear to be very old. The explanation is not obvious. Possibly, there is a certain category of comets that are relatively immune to aging, but this is just a hypothesis. Among the comets that have jumped far away from their original orbits, we find Hale–Bopp. After the in-depth studies of this comet, we know that its nucleus is very large, and this may be a good thing for survival. But it is impossible to tell how many revolutions it has actually completed.

If we try to trace the motion of this comet backward in time — even using the very best computers with the most cleverly constructed codes — we soon enough run into problems. The starting conditions are not exactly known, but there is an uncertainty that may amount to one year in the period of the preceding revolution. After tracing the comet exactly one revolution backward, we hence cannot know exactly where Jupiter was located. The perturbation that the comet received at its preceding passage is therefore estimated with a large uncertainty. If we then continue one more revolution backward, there is no way to tell at all where Jupiter was located as the comet passed, and any prediction about the perturbation received by the comet can be dismissed as completely unreliable.

This is an example of the role of chaos in the dynamics of the solar system. The equations that govern the motion have a unique solution, but only if the initial conditions are exactly known. Leaving the very least for chance to decide initially, it soon takes over completely. The position and velocity of the celestial body are in principle unpredictable. In this situation, one may often be helped by statistics. If one lets the computer calculate the evolution of millions of orbits, where each might be the real one at the initial moment, one may hope to obtain a picture of which evolutions are the most likely. But there is a danger involved, which comet Hale–Bopp may illustrate.

I have not made these calculations, but I'm still able to make statements about the results. If we follow millions of Hale–Bopp variants backward in time, we find that only a small minority ends up in the Oort cloud. The large majority, on the other hand, departs from the solar system. Does this really mean that Hale–Bopp likely came from the Galaxy and not from the Oort cloud? No, it does not. The outcome of such a calculation is the same, whether we trace the motions forward or backward, but there is a difference. In the future, everything that the computer predicts can happen, but not in the past. There are kinds of scenery that must be deemed unreasonable. Let us imagine a canary, which flies out of its cage and further into another room where a large window is wide open.

Now, imagine that we see a photograph of the bird during its flight, and we want to find out what was happening. As an experiment, we take a canary, place it where the photograph was taken and watch where it goes. We can be almost sure that it flies out of the window. It is much less probable that it flies to the cage and enters it. But this does not allow us to guess that the bird in the picture was coming from the street, where canaries are extremely rare.

The situation is similar with Hale–Bopp. We cannot totally exclude that it came from the Galaxy, but realistically, it has to be extremely uncommon that comets enter into the solar system at a speed as low as our calculations would indicate. On the other hand, we know that comets constantly arrive from the Oort cloud. If we instead look at the future of Hale–Bopp, this is very likely to imply the escape of the comet into the Galaxy, in case it physically survives until then. Beforehand, there is a small chance (as an estimate, about 10%) that the comet reaches an orbit so small that it is contained within the planetary system. It would then become a future analogue of Halley's comet.

There is a whole group of comets whose orbits show similarities with the orbit of Halley's comet. These are referred to as Halley-type comets. The most common opinion holds that they originate in random walks from the Oort cloud like I described earlier. But their current motions are far from the chaos that arises when the periods of revolution are much longer. We can safely make detailed predictions about the orbital evolutions during tens or hundreds of revolutions. Some of the comets have particularly complacent and regular behaviours due to resonances with Jupiter. The most common is the 1:6 resonance, where the period of revolution is on average six times that of Jupiter.

It is interesting that such a large group as the Halley-type comets have managed the long trip with their activity and brightness intact, in stark contrast to the large majority of comets. The reason is not clarified, but the properties of the orbits yield certain hints. While the newcomers from the Oort cloud have orbits with isotropic orientations, the Halley-types tend to have either rather small or remarkably high inclinations to the plane of planetary motion (Halley's comet belongs to the latter category). Such inclinations yield an increased chance of close encounters with Jupiter and thereby strong perturbations. The comets may thus have been lucky enough to complete the trip quickly, i.e. in a small number of revolutions. Another property is that the inner turning points of the orbits tend to be close to the Sun. This may increase the chances for the nuclei to clean their surfaces of the residual products that threaten to choke the activity.

Let me also mention another random walk, which appears to be performed by comets in a different region of the solar system. This concerns the inner core of the Oort cloud. The respective comets have fairly stable orbits, which still, slowly but surely, evolve over very long time scales. Comets may thus migrate from the true inner core into orbits with inner turning points in the Uranus–Neptune region. As a result, one of these planets may take control of the situation and start a random walk that results in a migration toward smaller binding energies and increasing orbital periods. The inner turning point is all the time far beyond the orbits of Jupiter and Saturn, so these giants have practically no influence. But as the period increases beyond a million years, the tidal effect of the Galaxy and passing stars start to work, and the comets can quickly reach the region around the Sun where they become visible as new comets from the Oort cloud.

It has been known for more than 10 years that this process may be responsible for a large fraction of the observed newcomers. Before then, it was believed that the inner core would not reveal itself by observed comets until the next comet shower occurs, probably in 100 million years or so, but now it seems that we do not have to wait at all. Since the perturbations by Uranus and Neptune are much smaller than those due to Jupiter, the random walk from the inner core involves much shorter jumps. Instead of jumping kangaroos, we can now think of usual frogs, which jump only a few decimetres at a time. These frogs run a much smaller risk of crossing the precipice, but they eventually meet their fate when they start crossing Jupiter's orbit by being transformed into kangaroos and performing the previously mentioned giant leap random walk.

4.3. Capture of Comets

I started my doctoral studies at the Stockholm Observatory at the beginning of the 1970s. My mentor was Prof. Lars-Olof Lodén. He was leading a project that was aimed to map the stellar population of the Milky Way within a region of the southern sky, and I began by assisting in this work. But Lodén advised me to seek new ways for my doctoral thesis. In particular, he thought of the renaissance that the somewhat dormant research field around the formation and evolution of the solar system was experiencing. It was obvious to all that American and Soviet spacecraft were exploring the Moon, Mars and Venus, but there was also a growing interest in theoretical studies, which would suit me as a mathematician.

Quite nearby — specifically, at the Royal Polytechnic — there was in fact a centre for such research. Its leader was the Nobel laureate in physics Hannes Alfvén. Thanks to the support of Lars-Olof, I came in contact with this research group. I had the opportunity to attend Alfvén's lectures and to personally meet him and some of his collaborators. They advised me on the topic of my thesis, and their proposed subject was the capture of comets. I followed the advice without hesitating and immediately plunged into the studies. On a beautiful day in May 1977, I defended my doctoral thesis, the title of which was exactly "The Capture of Comets".

The problem posed was to explain the existence of a large group of comets with the shortest known periods of revolution and remarkably low inclinations to the symmetry plane characterising the planetary system. This group is commonly called the Jupiter family, because the outer turning points of the orbits are situated close to Jupiter's orbit about five astronomical units from the Sun. In Chapter 1, I described a few Jupiter family members that have been visited by spacecraft, but at the beginning of the 1970s all these comets were quite anonymous or undiscovered.

It was clear that Jupiter exerts a large influence on the Jupiter family and often reshapes the orbits of these comets in connection with close encounters. Hence, this group must stay in contact with some other group of comets with different orbits, and Jupiter is mediating this contact. One goal of my investigation was to identify this group. But it was also known that Jupiter family members had disappeared in one way or another without Jupiter's involvement (see Section 1.2). Consequently, the other group must also serve as a reservoir. Fresh comets are gathered from there, and some die during their time in the Jupiter family. If this goes on for a very long time, Jupiter's orbital perturbations must provide a net inflow from the reservoir into the Jupiter family by means of so-called comet capture. My investigation would also elucidate what this means in a wider context.

Before I get to the thoughts that were spinning in my own head from the 1970s onward, I want to make a historic reflection. On Christmas Eve, 1740, in the town of Turku in present-day Finland, a boy was born, and this boy would grow up to become one of his time's most prominent scientists in mathematics and celestial mechanics. The boy's name was Anders Johan Lexell. He earned his Doctorate of Philosophy before his 20th birthday and a few years later moved to Uppsala, where he taught mathematics at the University. At that time, Catherine the Great ascended to the Russian throne, and she burned for the ideas of the Enlightenment. She quickly managed to recruit one of the greatest mathematicians of all times, Leonard Euler, to the Academy of St Petersburg. On Euler's initiative, an invitation was also sent to Lexell, who moved to St Petersburg and stayed there for the rest of his short life — he died at the age of 43.

In 1767, by the time that Lexell settled in Russia, something happened far from Earth, in the vicinity of Jupiter, of which no man had any idea, but which was later to be revealed by Lexell. It was a comet that for a long time had been moving in an orbit located between the outer part of the asteroid belt and Jupiter's orbit, which now encountered the planet in a very close fly-by. By the action of Jupiter's gravity, the comet was sent into a new orbit with its inner turning point somewhat inside the orbit of Venus. On its way there, it passed Earth's orbit, and Earth was there, right on time, for the meeting. At the end of June 1770, the comet was discovered with a telescope by Charles Messier as a modest object, but it brightened rapidly while approaching Earth. The smallest distance, achieved on July 1, 1770, was only six times the distance of the Moon — closer than any other approach by a known comet. The comet was then easily visible to the naked eye and, in addition, large and impressive due to its closeness.

Messier's comet of course enjoyed widespread attention. With a revolution period of only six years and an orbit that came very close to Earth, it made the scientists wonder why it had never been seen before. Lexell gave the answer to this question when he traced the comet's motion backward in time and was able to show that Jupiter's perturbations in 1767 had brought the comet from an orbit where it could not be observed into a new one that was nearly perfect. The comet has become known as Lexell's comet as one of the few exceptions from the rule that comets are named after their discoverers. Lexell shares this honour with, for instance, Edmund Halley.

Lexell also traced the comet's future motion and, once more, found something interesting. The period of revolution during the observations in 1770 was almost exactly one half of Jupiter's period. This is a very specific example of the previously mentioned 2:1 resonance. Lexell's comet became visible thanks to a close encounter with Jupiter in 1767, and after only two revolutions around the Sun (an interval of around 12 years), it was time to meet Jupiter again.

This fly-by in 1779 was like a mirror image of the first one. The comet was ejected into an orbit with its inner turning point near Jupiter's orbit, and the new period of revolution is likely very long. Many think that Lexell's comet is still pursuing its first revolution in the new orbit, and it has been suggested that it may never come back.

Lexell's comet was the first discovered member of the Jupiter family and has become a prototype for orbital changes triggered by close encounters with Jupiter. We may note in passing that there is a certain similarity between the comet's encounter in 1779 and the encounters that were planned by NASA in the 1970s to send the Pioneer and Voyager probes to the outskirts of the planetary system and beyond, using Jupiter's gravity as an energy source. When the spectacular encounter in 1779 was further investigated in the 19th century by scientists like Urbain Le Verrier, it was also discovered that the change of the comet orbit that occurred is extremely sensitive to the details of the encounter, such as the exact timing. Just like in the case of the long-period revenants from the Oort cloud, this is an example of chaotic dynamics, and the case of Lexell's comet was the first demonstration of the phenomenon.

As seen from the current perspective, when the Jupiter family has more than 300 members, it is remarkably often that such comets are discovered more or less directly after a close encounter with Jupiter, which changed their orbits so that they came much closer to the Sun — just like the case with comet Lexell. We have an example from the 19th century in comet Brooks 2, which was discovered in 1889, directly after an extremely close encounter with Jupiter in 1886, when the inner turning point of the orbit moved from 5.5 to two astronomical units from the Sun. From the 20th century, we have numerous examples, including comet Wild 2 (Section 1.5). In 1974, this comet came almost as close to Jupiter as Brooks 2, and the transfer of the inner turning point went from five to 1.5 astronomical units. The discovery followed at the first approach to the Sun in the new orbit in 1978.

There is probably an underlying reason, namely, that the comets become unusually bright and active in connection with such orbital changes. Afterwards, the activity wanes again since the surface layer of the nucleus gets enriched in rocks and organics, so that the evaporation of the ice and the leakage of the gas are hindered. The effect arising as the comet suddenly comes much closer to the Sun is that the earlier residuals are blown away by the increased gas pressure. Thus, the nuclei of newly captured comets ought to be extra fresh, and this has been an argument for choosing such comets as space mission targets.

Another important discovery concerning the mechanics of comet capture was made by Carl Gustav Jacob Jacobi and was published in 1836. In this case it was a question of pure mathematics. He investigated how a small body (for instance, a comet) is able to move in a space that is otherwise inhabited by only two large bodies (for instance, the Sun and Jupiter) if these have circular orbits around each other (for the Sun and Jupiter, this holds only as an approximation). Jacobi found that the small body has to move with a constant total energy if the motion is described in a rotating reference frame, where the two massive bodies are at rest. This constant total energy is called Jacobi's integral and may be used to approximately describe how comet orbits may evolve as a result of close encounters with Jupiter.

Let us consider one such application of Jacobi's integral. By simply assuming the comet orbit to be confined to Jupiter's orbital plane, and considering a diagram with the largest and smallest distances from the Sun characterising a comet orbit on the two axes, we can draw curves along which the orbital evolution has to proceed. This is in fact a good approximation, and the changes of real comet orbits — in particular, the jumps made in connection with close encounters — agree quite well with these evolutionary tracks. We therefore have a possibility to place hypotheses about the source of the Jupiter family by following the curves to larger distances. But unfortunately, this does not lead us very far.

If we follow the curves all the way until the largest distance is at least 20,000 astronomical units, we find orbits that characterise some of the newcomers from the Oort cloud. Their special property is that they have a very low inclination to the planetary orbits and an inner turning point close to Jupiter's orbit. The idea that the Jupiter family may stem from the Oort cloud is therefore tempting. When I started my doctoral studies, it had been known for a long time that such a transport cannot involve many cases where one single perturbation makes the comet realise the capture in one single jump. Thus, the issue was to study the effects of the random walks performed by all newcomers from the Oort cloud to see if these may in fact explain the Jupiter family.

This is the reason why, in the early 1970s, I strolled around with my head full of abstract thoughts around points that randomly jumped around along an axis in a diagram. In the spring of 1972, I had the opportunity to go to my first real conference abroad, which was held in Nice, on the topic of comets and other small bodies. I wrote to the organiser of the meeting and asked to participate — I introduced myself as a young student interested in the capture of comets. Imagine my amazement when the answer arrived with an invitation to give a review talk about this subject! I asked Lars-Olof Lodén for advice on what to do, and he answered without hesitation that I should go to Nice, give the talk and show them what I had learned. No sooner said than done, shortly afterwards I was standing on shivering legs in front of the learned audience and talking about things that I'm sure they already knew. In this audience, there were several world famous scientists, whose names I had repeatedly encountered in the literature that I had studied. The strange thing was their goodwill. I had applauses and no boos.

For me, the worst experience was that one of the authorities, an American professor Edgar Everhart, also gave a talk in which he showed the solution of the problem that I had introduced. He had performed simulations of the jumping points on a computer with the result that the capture of the Jupiter family works exactly in the way I sketched earlier (but which I had no idea about during my talk). Everybody seemed impressed, and I was wondering what I would now write my thesis about.

This problem was solved with the passing of time. I performed my works in the same field as Everhart and passed my exam

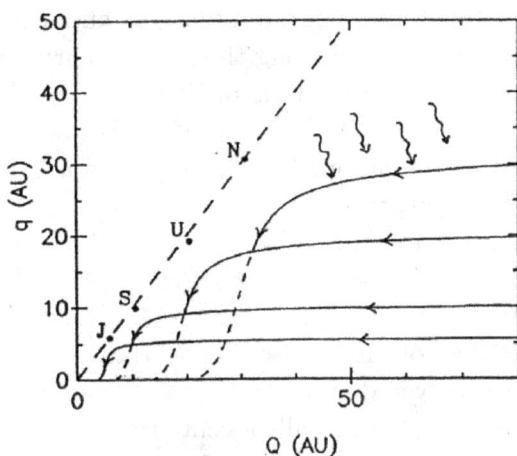

Fig. 4.3. The transfer routes followed by comets evolving between the transneptunian region and the Jupiter family, as explained in the main text. The x and y axes indicate the largest and smallest distances from the Sun, respectively. The orbits of the giant planets are shown close to a dashed line indicating circular orbits. The arrows show the direction of transfer when the comets are captured into the Jupiter family, and the curly arrows show the preceding evolution bringing transneptunian comets into contact with Neptune. Adapted from H. Rickman: *Origin and Evolution of Comets*, World Scientific, 2018.

five years later. After five more years, Everhart's hegemony would be broken. New ideas emerged, as I described in Section 2.3, and younger scientists (like Julio Fernández) made new simulations with results in conflict with Everhart's. For some time, it was hard to know who was right, but eventually, it became clear that the source of the Jupiter family was mainly among the transneptunians. Only a minor fraction would likely come from the Oort cloud (see Fig. 4.3).

During the transport of transneptunian comets into the Jupiter family, all the giant planets serve as a chain for passing along. Each of them would be able to transport comets along curves like those that describe Jupiter's effects — the shape is always the same in relation to the size of the planet's orbit. In practice, it works such that Neptune is the first planet to be visited, when comets from the Kuiper Belt or the scattered disk enter into orbits approaching or crossing the planetary orbit. When the comet encounters Neptune, it receives a gravitational kick that changes its orbit. The point describing the orbit in the diagram of largest and smallest distances from the Sun will then move along the evolutionary curve, on which it is situated. The motion can take place in both directions, but the most interesting evolution is the one that leads toward the Sun. The inner turning point of the orbit may then approach Uranus's orbit, and when this happens, the story is repeated with Uranus in the role of Neptune. Hence, Uranus may pass the comet on to Saturn, which in turn may finish the passing manoeuvres by bringing the comet into the vicinity of Jupiter.

One implication of this reasoning is that the celestial bodies that have been discovered in orbits situated in the region between the giant planets — the so-called centaurs — can be considered to represent a station on the road between the transneptunians and the Jupiter family. Sure enough, several of these have received designations as comets, since they have exhibited cometary activity. One of these is Chiron with a diameter of more than 200 km. It is possible that this centaur will perform as a guest of the Jupiter family in a distant future, which means that a gigantic comet may approach Earth's orbit (see Section 7.3).

In principle, all routes are open in this diagram, as long as they approximately follow the prescribed curves. But if one investigates where the Jupiter family comets end up in case they survive indefinitely, one finds that almost all of them will in due time be ejected into the Galaxy by Jupiter's action along the relevant curves. On the other hand, extremely few may originally have come along such a route. As we have seen, the main source is situated beyond Neptune's orbit. We may once again think of the poor canary, which came from its cage but, in all likelihood, will not return there but fly out in the street.

There is one remaining question concerning the Jupiter family: Which cage do these birds come from? Is it the Kuiper Belt or the scattered disk? It is unfortunately impossible to settle this issue by tracing the real comet orbits backwards. Close encounters with the giant planets render the orbital evolutions chaotic, so that in practice it is impossible to distinguish the true evolution from others that have nothing to do with the comet. To answer the posed question, one hence had to use a different approach. In a given time, on average, a comet in the scattered disk stands a much larger chance of encountering Neptune than a Kuiper Belt comet does. The only issue is how many comets there are in these two reservoirs. It is too complicated to observe small comet nuclei at such large distances, so we have to compare the numbers of objects with diameters of about 100 km or more and estimate the numbers of their tiny cousins from whatever we can infer about the respective size distributions. It has thus been found that the difference in numbers between the Kuiper Belt and the scattered disk is too small to compensate for the big advantage of the scattered disk, when it comes to the chance of capture. In conclusion, the Jupiter family mainly originates in the scattered disk.

Then, how many comets are actually hiding out there? I mentioned in Section 2.3 that the mass estimate is about 1% of Earth's mass for both the Kuiper Belt and the scattered disk. This is just a crude estimate, but it helps to provide part of the answer. Another part may be obtained by performing computer simulations of the capture process to see how efficient it is, but in addition, we have to know how many comets there are in the Jupiter family and for how long they survive.

4.4. Comets in Disguise

On Mt Palomar in southern California, a traditional observatory is situated. For many years, this was the host of the world's largest telescope — a reflector with a five-metre aperture. This remains and is still in use but is nowadays outclassed by many recent, much larger telescopes. Another main instrument at the Mt Palomar observatory is a large so-called Schmidt telescope, constructed to image large areas of the sky quickly and efficiently with minimal distortions. This telescope was used about 70 years ago to produce an atlas of the entire part of the sky that is visible from Mt Palomar.

During my time as PhD student, one of my tasks was to participate in public shows of the Saltsjöbaden Observatory. On such occasions, the Palomar Atlas was a safe card in case the weather was cloudy. I could unfold the spread where the large Andromeda Galaxy was situated, and success was guaranteed. Each time, the audience was equally impressed. But apart from this, among the crowd of inconspicuous dots, usually depicting stars or distant galaxies, these images also contained some previously unknown comets. Such a comet was discovered in November 1949 on a newly exposed Schmidt plate by Palomar astronomers Albert Wilson and Robert Harrington. This comet is named Wilson–Harrington and belongs to the Jupiter family.

However, these observations did not suffice to localise and find the comet on later occasions. It was eventually regarded as lost. Thirty years later, another astronomer named Eleanor Helin worked at the same observatory on a search for near-Earth asteroids (see Section 5.3). She then discovered an asteroid, which was designated as 1979 VA. At the time of writing, 40 more years have passed, and Helin's asteroid has been intensely scrutinised due to a discovery made in 1992, namely, that comet Wilson–Harrington and the asteroid 1979 VA are identical. But even so, one has not found even the least trace of comet activity in 1979 VA. It has been speculated that the comet tail on the Palomar Atlas was a chimaera caused by some mishap during the exposure, but there is now a consensus that a real comet was indeed observed.

Thus, we have before our eyes an example of a comet that has changed guise and become an asteroid. What can be the reason? It is probably a question of a normal comet nucleus, where the surface has become clogged to such an extent that no gas can pass. A couple of other Jupiter family comets are actually known to be almost "dead" with minimal activity. Their names are Arend–Rigaux and Neujmin 1. Since we know that such comets exist, it would be strange if there were not comets with even smaller activity — so small that we cannot even trace it. As a matter of fact, there are many asteroids that never showed any comet activity, but whose orbits are just like those of Jupiter family comets. There are also asteroids in orbits similar to those of Halley-type comets. The chances for usual asteroids to escape from the asteroid belt and be transferred into this kind of orbits are very small, and the spectra of these asteroids in cometary orbits look just like one would expect of extinct comet nuclei.

We are thus dealing with a phenomenon of common occurrence: comets lose their activity almost completely, but the nuclei persist, disguised as asteroids. It is not likely that they would have dried out completely and fallen asleep forever. The ice is likely still present somewhere at depth, and hence, there is also a possibility of rejuvenation. It cannot be excluded that Wilson–Harrington one day will wake up and once again behave like a real comet.

There are, apparently, comets that have woken up. The Slovak astronomer L'ubor Kresák investigated several cases where comets of the Jupiter family were discovered "too late". One example is comet Denning–Fujikawa, which was discovered by William Denning in 1881. If this had been anything near the same intrinsic brightness before 1881, it should definitely have been discovered earlier. Moreover, the behaviour of the comet after 1881 indicates that it may have undergone an outburst in connection with the discovery. In fact, it made ten returns from 1890 to 1969 without being observed and was rediscovered only in 1978 by Shigehisa Fujikawa. Thereafter, it was lost again until 2014. The observations from recent times show a faint comet with low activity, and it was probably like this before 1881 as well.

The idea of outbursts is natural, for outbursts have been observed in many comets. In 1892, Edwin Holmes discovered a similar comet. Like Denning, Holmes was an amateur astronomer and thus could only observe bright comets. Holmes's comet actually reached naked-eye visibility but then faded away. It returned twice as a very inconspicuous comet. Then it disappeared, and seven returns were missed until the recovery in 1964. After this, comet Holmes was observed at every apparition and always as inconspicuous. But on October 23, 2007, it flared up (see Fig. 4.4). In a few days, its brightness increased by a million times, and it was easily seen by the naked eye. It then fell back to its normal, bleak level, but in 2015, it presented another major outburst.

Comets are generally known to be capricious, and we see an important aspect here. The outbursts always arrive unexpectedly.

Fig. 4.4. Comet Holmes, imaged during its outburst in 2007. Credit: Iván Éder. License: Creative Commons Attribution-Share Alike 3.0 Unported.

No one ever predicted any of the observed outbursts. The reason is that we do not understand enough of the mechanisms that may be responsible. Our beloved comets are fooling us scientists in different ways — they disguise as asteroids, and they suddenly change between insignificance and magnificence. This is a disincentive for research, since it makes it difficult to judge how many comets the Jupiter family really contains, and thereby how many comets there should be in the scattered disk from where they come.

The changes in brightness and activity need not be due only to episodic outbursts, but they also seem to appear in the long term. We find one example with comet Encke (see Fig. 4.5). This has a unique position among the Jupiter family comets and is sometimes considered not to be a true member. This is because its orbit is extreme. The outer turning point is so far inside Jupiter's orbit that the comet never comes closer to the planet than about one astronomical unit. The orbit is hence very stable, and one immediately starts to wonder how the comet has ended up there.

Encke's comet has been known for much longer than 200 years. It is fairly easy to catch with binoculars, when the conditions are right, and sometimes it can be viewed with the naked eye. Since its discovery, it has completed more than 60 revolutions around the Sun,

Fig. 4.5. Comet Encke, imaged using the 0.91-metre Spacewatch telescope on Kitt Peak. Credit: NASA/JPL/J. Scotti.

and there is no indication that its brightness has waned. It rather appears with clockwork regularity. For how long can this have been going on? One important clue is that the orbit, which has been so stable for centuries, is experiencing a very slow evolution, which in the course of hundreds of thousands of years may make the comet fall into the Sun. The underlying reason is a particular resonance between the continuous turning of the comet's orbit and the corresponding turning of Saturn's orbit — in both cases, mainly caused by Jupiter's perturbations (see Section 2.2).

Most likely, comet Encke was captured as a normal member of the Jupiter family once in a remote past. But then something happened, which is experienced by extremely few comets. A very close encounter with Earth or Venus caused a radical change of the orbit so that this no longer approached the orbit of Jupiter. This marked the start of a very long period when nothing special happened. The orbit was totally stable, awaiting a new encounter with Earth or Venus, which might break the deadlock. The comet fell asleep and became a false asteroid. Eventually, there was an encounter, which diverted the comet onto a new track. The mentioned resonance became operational, and the slow march toward the Sun got started.

The false asteroid was gradually heated to higher and higher levels, until the pressure of the evaporating ice burst the protecting seal, and the comet came alive again. Some time later on, it became discovered by Earth's astronomers.

This story may seem complicated and far-fetched, but there is no doubt that it may capture the truth, and there does not seem to be any good alternative. In particular, the long-lasting sleep appears necessary, since if the comet had not been sleeping, the nucleus would have experienced fierce erosion by the long-lasting activity. In that case, Encke's comet would have started as a real giant — and, by a curious coincidence, we would now be witnessing it soon before its final demise.

4.5. Multiplication by Division

Some comets have truly extreme orbits, since the inner turning point is placed very close to the surface of the Sun. These are usually called sungrazers. When they approach the Sun, the nucleus is subject to such intense radiation and strong heating that lots of gas and dust are emitted. Therefore, the sungrazing comets often become enormously bright with very long tails. In ancient times, when people used to live under a virgin black night sky, such sungrazers were the most impressive spectacles that the sky could offer. Here, we find some of the most renowned comets of all times, and rumour about them may persist for hundreds of years.

The Great Comet of 1843 was such a comet (see Fig. 4.6). After its discovery in the month of February, it took just a few weeks before the comet approached the Sun. It was then so bright that it could be seen in full daylight, where it stood close to the Sun's disk on the light sky. The tail appeared in its full, impressive length as the comet had receded somewhat from the Sun and was easily watched low in the west in the evening sky. The next example is the Great September Comet of 1882. This one was also discovered a few weeks before its approach to the Sun, when it was shining much brighter than the Full Moon and was easily visible close to the Sun. It stayed under observation long thereafter and developed

Fig. 4.6. Contemporary drawing of the Great March Comet of 1843. Adapted from H. Rickman: *Origin and Evolution of Comets*, World Scientific, 2018.

a magnificent tail. Soon after its passage close to the Sun, the comet appeared to be split. At least five fragments came forth. Among the sungrazers of recent times, I can mention some of the most famous comets: Ikeya–Seki in 1965, Lovejoy in 2011 and ISON in 2013.

All the mentioned comets except ISON are closely related, but ISON sticks out in several ways. The orientation of its orbit differs markedly from those of the others. ISON is also a newcomer from the Oort cloud in contrast to the others. Last but not least, it behaved very differently. Instead of becoming extremely bright during its approach to the Sun, it faded away and seemed to disappear completely. The observations indicate that it split and quickly got consumed.

Splits have been frequently observed among the other comets, too, but there is nothing to indicate that the fragments would not return. We deal with periods of revolution between 500 and 1,000 years. Such orbits are typical of old revenants, which have completed many revolutions around the Sun, since once upon a time they came from the Oort cloud. The difference in behaviour between ISON and the others may therefore give a hint about the physical evolution of the comets. Among the new comets, there are all variants, from very weak to really strong, but among the old comets, only the strong ones are left.

Let us now have a closer look at the old sungrazers. The natural idea that all of them are returning fragments from earlier splits dates far back in time. Already in 1888, the German astronomer Heinrich Kreutz found evidence in favour of this hypothesis, and the whole group of related sungrazers has been named the Kreutz group in his honour. A great comet that was seen in the year 1106 has been found to fit very well as the parent comet of the Great September Comet in 1882 and comet Ikeya–Seki in 1965. However, to put together the puzzle of the entire group, one has to go at least one more revolution backwards. According to one unverified idea, the original mother might be a comet of antiquity, which Aristotle observed in the 4th century.

No matter at which time the original mother of the Kreutz group made its fatal approach to the Sun, its nucleus must have been unusually large. It could thus host the material of several fragments, each of which gave rise to large and bright comets, and thereafter a series of large and bright comets in the following generation as well. Of course, this cannot go on forever. When the fragments become too small, they are unable to survive but are dispersed and annihilated by the solar heat, if they do not actually rush into the Sun.

In fact, this very process has been very well documented. One of the most important space telescopes named *Solar and Heliospheric Observatory* (SOHO) was launched in 1995 and is still in use through a collaboration of NASA and ESA. As a main part of the observations, SOHO has continuously imaged the inner regions of the solar corona, and more than 3,000 comets have been discovered on these images. Taken together, these completely dominate the list of all comets that have been observed in historic times. Even though the SOHO observations are normally insufficient to determine accurate orbits, it has been concluded that the large majority belong to the Kreutz group.

Here, it is a question of miniature comets — actually far too small to deserve a place on the same list as the usual comets. What we see is the small rubbish that is produced in connection with the large splits. SOHO has observed the final destruction of this rubbish for more than 20 years, which may be seen as a long time. But yet, we cannot know if the whole orbit with about 800 years' revolution period is full of this waste, or if we just happen to watch a clump of material currently situated in the innermost part of the orbit.

Faced with this story about the origin of the Kreutz group, it is natural to wonder how the mother comet entered into the fatal orbit along which it nearly collided with the Sun. This orbit must have arisen as a result of an evolution from earlier, harmless orbits. Now, the previously mentioned processes that provide us with new comets from the Oort cloud do not work in the case of the Kreutz group sungrazers. These comets have too small orbits to feel the effects of the Galaxy or passing stars. The evolution is instead governed by the giant planets — in particular, Jupiter.

Some comets arrive from the Oort cloud on orbits that are inclined at nearly right angles to the planetary orbits. The road toward Halley-like orbits with periods of about 100 years then becomes troubled. The very elongated orbits, whose inner turning points are originally situated at distances from the Sun similar to that of Earth, may gradually be reshaped by Jupiter's gravity so that these points are pushed even closer to the Sun, while the inclination decreases. This evolution is eventually reversed and in the long run proceeds in cycles. The oscillations in question are usually referred to as the Kozai effect after the Japanese astronomer Yoshihide Kozai. But if the comet approaches the Sun too closely, so that the nucleus is broken apart by tidal forces, the effect is stalled by the disappearance of the comet.

It is obvious that the split of the original mother of the Kreutz group has given us many more comets than we would otherwise have had. Of course, this is the case with all the splits — also those that sporadically affect comets of all kinds without any influence of tidal forces. We saw one example involving comets Neujmin 3 and

Van Biesbroeck in Section 1.6. In that case, the Jupiter family has been concerned, and the issue is how large the dark figure may be. Similar events that may possibly have occurred many centuries or thousands of years ago would be impossible to trace owing to the chaotic orbital evolutions of the Jupiter family comets.

It is of interest to speculate about what may happen if Chiron one day is captured into the Jupiter family (see Section 4.3). Quite likely, very large comet nuclei do not split as easily as small ones do, but we cannot know for sure if they are immune. Hence, there is a possibility that, on rare occasions, the Jupiter family is enriched by a giant of Chiron's type and turns into a giant family, as this guest splits. Let me note that, in principle, every centaur may be captured into the Jupiter family some time during its history, and Chiron is not the largest one. The current record is held by Chariklo, whose diameter is 260 km. For Earth, the consequence of such temporary giant families would be time intervals with a drastically increased frequency of impacts due to comet fragments of all sizes.

As a final remark, the comets may, like paramecia, multiply by division. This should perhaps be taken into consideration when one estimates the number of comets in the Jupiter family and their lifetimes. But once again, we stand before a very difficult problem, which has barely started to be attacked. The comets lead a complicated life, and their ageing and passing depend on how they are internally characterised as well as how their orbits evolve. It is still the case that we sometimes stand bewildered at the changes they exhibit.

Chapter 5

Space Projectiles: A Hazard to Humanity

Disaster movies play a prominent role in contemporary American popular culture. These paint horrific visions of natural disasters that threaten to exterminate humanity and ravage society until, typically at the last moment, they are averted by Bruce Willis or some other action hero. Among the most popular villains in these movies, we find comets and asteroids on collision course with Earth.

Is all this pure fantasy, or is there any real background? Yes, there is such a background, and it has to be described in a factual and correct way. Many years ago, while seated in my office, from time to time I was called on the phone by a man who suffered from bad anguish due to this cosmic collision risk. These conversations used to occur when he would hear of an imminent close encounter via TV or some newspaper. He would call me in order for me, as an expert, to appease him and give him my word of honour that there was no risk of collision. I believe that many others had similar problems and quietly suffered awful agony before the predicted dates of the world's demise.

Let us therefore scrutinise the issue of the cosmic collision risk in all its aspects. We shall see which collisions have occurred, what damage they have done, how often they are expected to arrive, and finally, how we can decide if a given asteroid poses an actual risk and, in such a case, what we can do to defuse it.

5.1. Our Real Experience

A meteorite fall would doubtless be deadly if the meteorite hit one's head. But meteorite falls are so unusual — and the total area of all the heads in the world so small — that no such case is known. It is Earth's atmosphere that protects us by screening out almost every rock that arrives from space.

The danger is instead due to the very largest meteorites whose explosions in elevated layers send dangerous shock waves to the ground. When I now speak of danger, I mean meteorite falls large enough to be able to kill a large number of people and frequent enough to normally occur somewhere on Earth during an average human lifetime. In this case, too, the real mortality rate is so far equal to zero, since the bursts have generally taken place over poorly inhabited areas, but in at least one case, the call was close (see Fig. 5.1).

In the morning of February 15, 2013, the inhabitants of the Russian city of a million, Chelyabinsk, a little Southeast of the

Fig. 5.1. The Chelyabinsk superbolide explosion, registered from a car by a dashboard camera. Associated Press.

Ural Mountains, could see the sky enlightened by something called a superbolide. It was brighter than the Sun and came from a terrible explosion at an elevation of almost 30 km. The energy is estimated at $1^1/_2$ megaton of TNT. The cause was a meteoroid with a diameter of about 20 meters, which hit Earth at a speed of about 20 km/s. This was reported in detail by news media all over the world, and many will likely recall the videos of the bolide that were registered from cars in the morning traffic.

The shock wave caused widespread damage to the buildings of Chelyabinsk. Windows were cracked or smashed in more than 40% of the residential buildings. Straight beneath the fireball track, people were felled to the ground by the pressure. Many cases of sudden sunburn from the ultraviolet radiation of the bolide were witnessed by people who were outdoors on the occasion. It is estimated that this explosion could have killed many people if its energy had been somewhat higher or the elevation of the explosion had been a little lower.

The meteoroid was already split into smaller parts when the explosion occurred. All in all, less than one per mille of the original mass is thought to have reached the ground in the form of meteorites, while the rest largely evaporated or pulverised into dust. The largest recovered meteorite was revealed by a seven meters wide hole in the 70 cm thick ice on a lake named Chebarkul. Indeed, one was able to find and recover a stony meteorite of more than half a tonne from the bottom of the lake beneath the hole.

Among all the events of this kind, which are well documented and placed beyond doubt, there is only one that surpasses the one at Chelyabinsk. The day when this occurred was June 30, 1908. This time, too, the site was located in the trans-Ural part of the Russian empire, though considerably further to the East. The place was far from any town or larger village, so the locality is named after the river Podkamennaya Tunguska. The area was inhabited by Evenks and Russian settlers.

It was in the wee hours of dawn that these saw the sky lit up by an immensely bright light, after which there was a lightning and sounds like gunfire. Those who were closest to the burst were felled to the ground by the pressure wave, and windows were smashed hundreds of kilometres away. It is estimated that the explosion occurred at an altitude of 7–8 km. Due to its enormous energy, the shock wave rocked the ground so that signals were received by seismographs all across the Eurasian land mass. The strength of this earthquake is estimated at Richter magnitude 5, and the explosion energy corresponds to about 10 Mt TNT.

The Tunguska event also had other effects. In Scandinavia, attention was drawn to the light nights that characterised the time shortly afterwards. Newspapers could be read and photographs taken outdoors at times of day when this was normally impossible. This was caused by dust dispersed by the explosion into high atmospheric layers. As a result, sunlight was scattered to places on the ground that, normally, it would not have reached. This dust also reduced the transparency of the atmosphere, which was ascertained by astronomical measurements.

Almost 20 years would pass before the site was reached by a scientific expedition led by the Russian mineralogist Leonid Kulik. When they arrived there, the scientists were met by several surprises (see Fig. 5.2). First, there was no crater, which means that no large

Fig. 5.2. Devastated forest area near the epicentre of the Tunguska 1908 atmospheric explosion. Photo by Vokrug Sveta, member of the 1929 expedition to Podkamennaya Tunguska, published 1931. In the public domain.

part of the meteoroid reached the ground. But no usual meteorites could be seen either. Instead, what was plainly visible was the devastation after the explosion. The forest was felled with trees lying with their tops radially out from the epicentre. At the very centre, stripped trunks were still standing. During later investigations, the devastated forest area was measured at $2,150 \, \text{km}^2$.

It is fairly obvious that no human being would have survived inside this area. But, according to most sources, no casualties were incurred. However, one can easily imagine what might have happened if the explosion had occurred above a major city. The devastated area at Tunguska is similar to that of Paris, including suburbs. The difference between the consequences of Tunguska and Chelyabinsk is due not only to the former's much larger explosive power but also to the fact that the Tunguska meteoroid penetrated much deeper into the atmosphere before the explosion took place.

There has been an extensive debate about what this meteoroid consisted of and which physical structure it had. A few decades ago, many were of the opinion that the impactor was a small comet. The fragility of comets could explain the lack of meteorites and the total disintegration of the object in the atmosphere. In addition, the date — June 30 — falls during the annually recurring meteor shower beta Taurids, which is known for its relation to Encke's comet. Thus, the issue is whether it was a small piece of Encke's comet that exploded over Podkamennaya Tunguska.

However, this hypothesis has been doubted in more recent years. The main argument is that the very fragility of comet nuclei is difficult to reconcile with the fact that the body descended to less than 10 km height before bursting. Nowadays, the majority opinion is that the body consisted of rocks like the ordinary chondrites. Its diameter is estimated at about 50 metres. However, I want to stress that the last word may not have been said about the issue of the object's composition and structure, as well as its possible relation to Encke's comet.

Considering the hazard to life and property posed by atmospheric explosions like the Tunguska event, it would be of value to know how often these occur. It is unfortunately impossible to give an accurate figure for this frequency based on real experience. Our experience is simply too small. The Tunguska event seems to be in a class of its own, considering the time since the beginning of the 20th century, but if we go much further back, it becomes much harder to decide what may actually have occurred.

One event purported to have taken place in Qingyang, China, in 1490 may serve as an example. Preserved scriptures from the time in question tell of a rain of stones that fell from the sky and killed more than 10,000 people. But no details are reported, and no meteorites have been found. To elucidate what really happened may be a task for historians rather than astronomers.

However, we have good means to determine the frequency of other events — of both smaller and larger extent. High altitude airbursts with an explosive power of several kilotonnes of TNT — thus, comparable to the Hiroshima bomb — occur often and are carefully surveyed by the detector systems of the superpowers. An important reason is the wish to avoid the risk to "push the button" by mistake. From this source — in particular from the USA — there are public statistics of great value. On the other hand, we also have the gigantic impacts that create large craters. I shall describe their frequency in Section 5.3, but for now we can state that this is fairly well known, even though no such impact is thought to have occurred during the whole time that homo sapiens have lived on Earth.

Taking these two frequency estimates and interpolating between them, one may arrive at a rough estimate of the frequency of Tunguska-like events. The result is in a sense surprising. In fact, we should expect only one such event in a thousand years, so it may seem remarkable that the last one occurred only a hundred years ago. A similar comparison for the Chelyabinsk event shows the same thing. In this case, several hundred years should on average pass from one such event to the next, so it is indeed remarkable that the last one happened less than 10 years ago.

Some scientists have claimed to see an indication that we are currently experiencing a period of unusually high crash frequency of bodies with 20–60 m diameter into Earth. We cannot say for sure if that is the case, but if it is, a possible explanation is readily available.

In case small Earth-crossing asteroids or comet nuclei break apart due to collisions or for other reasons, one can imagine broad streams of fragments to form along their orbits in a similar way as the previously mentioned meteor streams originate. Earth may then be hit with a higher frequency than otherwise during an epoch when such a stream is actually pierced by Earth's orbit. However, there is no apparent association between the orbits of the Chelyabinsk and Tunguska impactors.

5.2. The Traces of Collisions

When I read about the solar system in books before my doctoral studies, texts concerning the lunar craters reported that their origin was unclear. Some claimed that these had arisen through impacts by asteroids and comets, while others preferred to see them as volcanic formations. Let me now describe the background.

Already Galileo saw the lunar craters through his telescope, but at that time the nature of comets was unknown, and no asteroid had been discovered yet. It was only in 1898 that an astronomer made the first discovery of an asteroid whose orbit came close to that of Earth. This was named Eros (see Fig. 5.3).

After close-up studies by the space probe NEAR Shoemaker, we now know that its average diameter is 17 km. But these objects are generally faint and difficult to observe, and new discoveries were initially made at a very slow pace. In 1930, only three more asteroids were known with the same kind of orbits. All of these had their inner turning points outside Earth's orbit. Neither of them therefore posed any immediate threat of crashing into Earth.

The first asteroid whose orbit stretched both inside and outside Earth's orbit was discovered in 1932 and was named Apollo. As time passed, more of these were discovered, but the number stayed low for a long time. Moreover, there still was no consensus about the innermost nature of comets. It is therefore no surprise that the debate on the nature of lunar craters lingered on until fairly recently. But, in any case, the realisation that there are asteroids that pose a non-negligible impact risk to Earth and the Moon raised a certain

Fig. 5.3. Image mosaic featuring asteroid Eros as observed from the NEAR Shoemaker spacecraft. The asteroid is seen from a direction near the North Pole. Credit: NASA/JPL/JHUAPL.

interest among geologists to find out if we could see traces of such impacts among terrestrial craters.

This led to a scientific progress that in my opinion is impressive. A new scientific branch, called impact geology, has been born out of nothing. This was shaped in the 1950s by studies of a few craters — obvious as well as suspected. In the USA, most of the interest was focused on the Barringer crater in northern Arizona. This is surrounded by a field of iron meteorite finds, and the Americans used to call it Meteor Crater (see Fig. 5.4).

The crater was claimed in 1903 by the Barringer family, who planned to mine iron from a giant meteorite buried under the crater floor. Drills were started but yielded nothing of value. The rusty, old equipment still stands in the crater. The fact that no big meteorite was found gave rise to some scepticism concerning the impact origin of the crater — maybe the surrounding meteorite field was just a coincidence.

Fig. 5.4. The Barringer crater in northern Arizona as seen from the North rim. Credit: Grahampurse. License: Creative Commons Attribution-Share Alike 4.0 International.

However, another legitimate suspicion was that a large enough celestial body made largely of iron might pass the atmosphere practically undamaged. The explosion arising when such an object hits the ground with a speed of tens of kilometres per second would be comparable to nuclear bombs of Megaton class. Moreover, some lessons had been learned from the results of the small-scale nuclear tests that were performed during the 1950s in the USA and the Soviet Union. It thus appeared likely that almost the whole body would be vaporised together with an even larger mass of the underlying bedrock.

Hence, there is no disagreement between the impact hypothesis and the lack of a buried body. But the hypothesis still needs factual support, and this is where the work of geologists comes in. Using theoretical models, it is possible to predict the results of an impact, i.e. the effects to be observed afterwards in the surrounding bedrock. If geological field studies confirm what is predicted and nothing else, the impact may be taken for granted. In such a way, already 60 years ago, the geologist Eugene Shoemaker managed to show convincingly that the Barringer crater is an impact structure.

Over time, definite evidence would also emerge with the discovery in rocks at suspected impact sites of minerals that can only be formed as a result of impacts at cosmic speed. The critical factor is pressures in the gigapascal range, which occur naturally in Earth's deep interior but not near its surface. Such pressures arise during an impact when the shock wave passes through the bedrock. A few well-known examples are provided by the minerals stishovite and coesite, which represent different forms of shocked quartz.

The time of formation of the Barringer crater has also been determined through radioactive dating of the shocked rock. This impact occurred about 50,000 years ago. From the size of the crater (its diameter is 1.2 km, and the depth is 170 m), one can also estimate the diameter of the celestial body at about 50 metres. The crater still appears very fresh, since the climate has remained dry and the natural erosion due to precipitation and vegetation has been absent.

If the impacting body had consisted of ice or the kind of rock that most meteorites are made of, it most likely would not have reached the ground, and no large crater would have been formed. Many such atmospheric explosions must have occurred during the last 50,000 years. Metallic bodies have probably also impacted the Earth after the formation of the Barringer crater, but these fell into the ocean or exploded at places where erosion or covering has erased the traces.

But 50,000 years is just a moment in the history of Earth. Taking the perspective of many million years, much more impacts must have occurred, and some of these must have made imprints that were gigantic in comparison to the little Barringer crater. Can these giant craters still be identified? Yes, this is possible even though the traces have been partially erased or modified, so that the identification is impeded. However, there are large parts of Earth's surface where not even the largest craters have withstood the geologic evolution — this concerns, for instance, mountain range folding and gigantic lava flows.

It is thus on the most tranquil parts of Earth's crust that the largest and oldest impact structures are most easily found. One example is found in my native region, i.e. the Nordic countries. As if by luck, this region has also hosted many prominent impact geologists. Thus, a large number of local impact craters have been evidenced. In the southeast of Småland, one of the southern provinces of Sweden, there is a round lake named Mien spanning an area of $20\,\mathrm{km}^2$, and in the middle of Mien there is an island named Ramsö. One may naturally imagine this as a water-filled impact crater with a central peak that emerges above the surface of the lake. Ideas with this connection were brought forward more than 100 years ago and received strong support in the 1960s, when shocked quartz was found in the area. It took until the 1970s, though, before the geologists who advocated a volcanic origin had to give in. The age of this crater has been estimated at 120 million years.

In any case, number one among the Swedish impact craters is the Siljan ring. This is one of the largest in Europe, and it was early suspected to be an impact structure. Starting from the 1950s, geologic investigations were carried out in order to settle the issue. By this means, the impact hypothesis has eventually been confirmed. I have been fortunate enough to see different kinds of evidence at the very place, and these are indeed impressive. There are, for instance, places like quarries, where one can watch the bedrock exposed in time profile and note that the geologic layer sequence is inverted. The deepest layers are on top, and the most superficial ones are at the bottom. During the formation of the crater, the bedrock was flopped into the air and landed upside down just beyond the crater rim.

I had a particularly strong impression when I was taken to a remote place in the big forest and grabbed pieces of granite in my hand. The place was near the centre of the crater, and the shock wave had crushed the granite so that I could crumble it between my fingers. At that moment, I realised better than ever, what asteroids and comets can do to our Earth. It is also impressive to imagine what the crater looked like just after it was formed about 370 million years ago. Its diameter was about 55 km. Its rim was a circular mountain chain with heights of a few thousand metres, and the peak at the centre rose as high. But all this has been eroded away during the following time. What is left is a ring of lowlands that originally were located just inside the rim. This is easily spotted on a map as a series of big lakes, namely, Siljan, the Orsa Lake, Skattungen and Lake Ore (see Fig. 5.5). On a usual road atlas, it can be followed along the county roads 296 and 301, the national road 70, and European road E45.

If we count the number of confirmed impact craters around the globe, the result is about 200. A remarkable fraction is situated in or near the Nordic countries. A few examples are Lake Lockne south of Östersund in Sweden; Söderfjärden just south of Vaasa in Finland; and the Gardnos crater in Hallingdal northwest of Oslo. The reason for this concentration, of course, is not that asteroids and comets have anything against Scandinavia, but we conclude that craters elsewhere have been erased to a larger extent or still remain to be discovered.

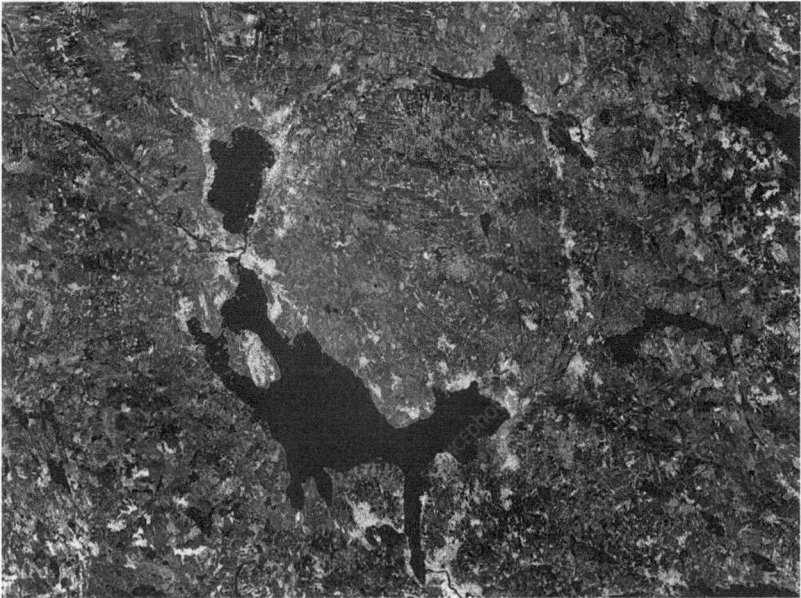

Fig. 5.5. Satellite photo of the Siljan ring in the Swedish province Dalarna. Several lakes stand out along the perimeter of the ring. The largest one is Siljan, Sweden's seventh largest lake. Credit: PLANETOBSERVER/Science Photo Library.

Let us now return to the lunar craters. The old issue of their origin was finally settled by the finds from the NASA Apollo landings and the Soviet Luna probes in the early 1970s. The lunar surface proved to be covered by a loosely aggregated layer of rocks and gravel, which was fairly comfortable for the astronauts to walk on. The constitution of this layer was relatively similar between the different landing sites. Rocks of different origins were randomly agglomerated like the contents of a well-mixed bowl of candies. The geologists call this a breccia. Some pieces consisted of dark basalt and others of light anorthite. The basalt dominates in the lunar maria and the anorthite in the highlands. The fact that these components appear mixed together at so many places indicates that the bedrock is crushed everywhere and pieces have been flung around over large distances.

This, together with the verification of the melt and shock effects to be expected from impacts, provides sufficient proof that the lunar craters are almost entirely impact craters. That this is the case was certainly an important discovery, but it also helped to convince the sceptics of the extraterrestrial origin of terrestrial craters. Whatever the case may be with the space projectiles — if they hit the Moon, they must also hit Earth, and this happens even more often because of Earth's larger cross-section and stronger gravity.

But what is more: if the Moon and Earth get hit, it would be strange if the other planets would not also be impacted. On the images of Mars and Mercury from space probes at close distance, we indeed see crater landscapes like that of the lunar highlands. Local differences inform about the history of the celestial bodies. For instance, on the Moon the craters are much fewer in the maria than in the highlands, and the northern hemisphere of Mars is much less cratered than the southern one. Craters are very few on Venus, but they do exist. I shall return to this in Section 5.5. In any case, the matter is settled. Over time, all the planets have been exposed to a bombardment by asteroids and comets, which has scarred their surfaces. This bombardment is still ongoing.

5.3. The Impact Risk and the Near-Earth Asteroids

Sometimes, the bombardment of the planets is used to date their surfaces. In a relative sense, this is simple: the more crowded by craters the surface is, the longer is the time of its exposure to the bombardment, and the older the surface is. Therefore, we can directly tell that the lunar maria must be younger than the highlands. Absolute numbers for the ages of the surfaces will be discussed in Section 5.5, but we can forecast that the age of the lunar maria is somewhat more than 3 billion years. This was made clear during the exploration made at the beginning of the 1970s. With the aid of crater formation theory, one could also estimate the sizes of the celestial bodies that had caused the observed craters.

One interesting conclusion was that the number of craters can only be explained if the average number of bodies able to collide with the Moon in the past has been much larger than the number of such objects that was known around 1970. The most natural interpretation is that the current time is typical of the average, but that the large majority of the celestial bodies — kilometre-sized comet nuclei and asteroids — remain to be discovered. It seemed too unlikely that an immense number of comets would exist without being discovered, so the interest was instead focused on small asteroids of the types of Eros or Apollo. These are referred to as *Near-Earth Asteroids*.

At that time, the search for the undiscovered asteroids had a dual aim. In addition to the defensive search for Near-Earth Asteroids, there also was a focus on the main belt between the orbits of Mars and Jupiter. In fact, it was clear that the meteorites originate from these asteroids and that both the meteorites and their parent bodies carry useful information about the origin of the solar system. The number of asteroids on record at the start of the 1970s was less than 2,000, but it was realised that the main belt contained enormous quantities of as yet undiscovered members. The exploration of these might open up new routes to learning how our solar system was born.

At the same time, I was a newly admitted PhD student at Stockholm Observatory, and I got a task that was extremely interesting. I already mentioned my contacts with Hannes Alfvén and his team at the Royal Polytechnic. It so happened that Alfvén received financial support for a Nobel Symposium, which would be focussed on the origin of the solar system and the ideas that he himself had introduced. A few dozens of the world's most celebrated scientists were invited and came to the meeting, which was held at the Grand Hotel in Saltsjöbaden (very close to the Observatory) in the summer of 1971. My specific duties were, on the one hand, to receive the honoured guests at Arlanda airport and provide necessary assistance, and on the other hand, to sit in the back of the stage during the sessions and run the tape recorder, which would document every single word of the discussions.

Later, I was also asked to help edit the proceedings, but I especially recall the good advice I got from some of the participants. The astronomer Tom Gehrels pointed out the need to discover more asteroids, and this made a deep impression on me. When I talked about this with my supervisor, Lars–Olof Lodén, we concluded that one ought to be able to perform the search through international collaboration. On his initiative, I wrote to all the most prominent scientists to hear their opinions. The result was not as nice as I had hoped. A few of the answers sounded like pure reprimands, and I guess that they were affronted by a youngster from Sweden trying to meddle with their work.

At any rate, by that time the first systematic search projects were coming online in the USA. These used Schmidt telescopes with large fields of view, and glass plates of the same stellar field taken at somewhat different times were placed in a so-called blink comparator. Here, the two images were seen reciprocally. While the stars stood still, one could sometimes see a light spot jumping to and fro, revealing an asteroid. In this way, it was easy to identify Near-Earth Asteroids, which due to their closeness move rapidly across the sky and thus jumped longer than normal.

These projects gave the expected return. The number of known Near-Earth Asteroids grew considerably faster than before. One of

Fig. 5.6. Eugene M. Shoemaker (1928–1997), American geologist and astronomer. This picture shows Shoemaker at work with a stereoscopic microscope. Credit: US Geological Survey.

the first to embark on this activity was Eugene Shoemaker — the geologist who had made essential contributions to the identification of terrestrial impact craters (see Fig. 5.6). Now he also became an astronomer, looking for the projectiles among the stars of the sky. Somewhat later, Tom Gehrels developed a very successful project called Spacewatch at Kitt Peak National Observatory in Arizona — one of the first to be largely computerised and to work with electronic images instead of photographic plates. This was inaugurated in 1980 and contributed to a gradually increasing discovery rate for Near-Earth Asteroids.

However, the most notorious event in 1980 within impact research was not the inauguration of Spacewatch but something completely different. A scientific paper by Nobel laureate Luis Alvarez, his son Walter Alvarez and two other scientists was published in the journal *Science* and directly made a sensation. The title was "Extraterrestrial Cause for the Cretaceous–Tertiary Extinction".

The boundary between the geologic periods Cretaceous and Tertiary — the K/T boundary — occurred 65 million years ago.

It was then that the dinosaurs became extinct, along with many other species. About half of all the animal species living at the end of the cretaceous suddenly disappeared. Such an event is called a mass extinction, and the one at the cretaceous–tertiary boundary is one of the five most extensive mass extinctions that palaeontologists have identified since the Cambrian explosion about 540 million years ago, when complex life forms took control of Earth's surface. I shall return to this in Section 7.3.

Formally, the term tertiary has now been replaced by paleogene. One should hence say cretaceous–paleogene instead of cretaceous–tertiary, but the old term has become so engraved in the language that it still survives. The reason has a lot to do with the excitement raised by the Alvarez *et al.* paper and the long-standing debate that followed. The discovery described in the paper was built on studies of a lithified clay layer that characterises the K/T boundary around the world (see Fig. 5.7).

This is exposed at many places, and one easily accessible site is Stevns Klint south of Copenhagen. In the mentioned clay layer,

Fig. 5.7. Detail of a sedimentary rock from Wyoming, USA, exhibited at the San Diego Natural History Museum. The light coloured streak running horizontally across the image is the claystone layer formed at the K/T impact 65 million years ago. Photo Credit: Eurico Zimbres. License: Creative Commons Attribution-Share Alike 3.0 Unported.

Alvarez and his team found an enrichment of the heavy metal iridium, which appeared similarly independent of which site they examined.

Compared to other chemical elements, iridium is not remarkably rare in the Sun or in the chondritic meteorites, but it is exceedingly rare in Earth's crust. This is thought to be due to the chemical separation that took place inside planet Earth as a consequence of an extreme heating very early in its history. The iron sank to the bottom, forming the iron core at the centre of Earth. A series of other elements, among which we find iridium, followed the iron to nearly 100%, and these are thus extremely rare in the crust. But if a gigantic meteorite (or a comet) impacts, its iridium enters into the blanket of ejecta that surrounds the entire Earth. The idea is that the K/T clay layer is just such a blanket from a giant impact. It then becomes logical to imagine that the mass extinction at precisely the same time was not a chance occurrence but was caused by the impact.

This brought a new dimension to the impact research. Thus far, this had only concerned craters, but now it concerned life as well. The debate has undulated back and forth around the issue of how close the connection is between giant impacts and mass extinctions. In the case of the K/T boundary, it is hardly possible to deny that the connection exists, but there is no consensus about the details. In general, it appears that giant impacts offer just one of several explanations for mass extinctions, and it also seems uncertain whether every such impact automatically sets an imprint on the evolution of life.

This does not prevent that extensive work was devoted to understanding the mechanisms behind the global hazard of the impacts. There are analogues between this research and the one aiming to understand the effects of nuclear war. The term "nuclear winter" stands for the extreme deterioration of climate, which would be a result of all the dust that the bombs would spread into the atmosphere. A giant impact would trigger the same effects, even worse though.

What Alvarez *et al.* put forward was not an absolute proof of a giant impact but rather a well-founded hypothesis. However, the proof was to come. Already in 1978, geophysicists working for the Mexican state oil company Pemex had drawn attention to a circular shaped magnetic anomaly at the northern part of the Yucatan peninsula. Earlier still, a map of local gravity anomalies had indicated a considerably larger, circular structure at the same place. There were suspicions of a very large impact crater, but the measurements were initially concealed due to the important economic interest coupled with the oil. It thus took almost a decade from the sensational Science paper until the geologic evidence for the gigantic crater and its age of 65 million years could be published. The strength of evidence provided by the crater concerning the impact origin of the K/T layer has been likened to a smoking gun in a murder investigation. The crater is named after the small town Chicxulub on the northern coast of Yucatan. Its outer diameter is 240 km, but the diameter of the initial crater is estimated at 150 km.

As the end of the 1980s was drawing near, almost 100 Near-Earth Asteroids with diameters of at least one kilometre were discovered. This was enough to estimate the total number including all the undiscovered ones with a reasonable accuracy. Judging from the average collision frequency with Earth for such objects, it was also quite clear that about one such collision is expected to occur within a million years. This became interesting indeed, when combined with rough guesses about the consequences of such an impact, for instance, in case any of the undiscovered asteroids would hit Earth in a few years' time. Owing to the previously mentioned climatic damage together with other atrocities, Earth's human population would likely be significantly decimated, and the mortality rate was estimated at a billion people.

Two American asteroid experts, Clark Chapman and David Morrison, wrote a book with the title *Cosmic Catastrophes*, where they discussed different aspects of the impact hazard. According to their judgement, the greatest hazard to humanity arises from the kilometre-sized asteroids owing to a combination of a non-negligible impact risk and globally disastrous consequences. They claimed that this hazard merits serious attention and proposed measures to be taken to better understand the extent of the risk and, also, to be able to mitigate it, should that be necessary. A striking illustration of the issue is that the average mortality rate from asteroid impacts over a period of a million years is about 1,000 per year, which is comparable to the corresponding estimate for air crashes.

This comparison is not quite fair, because we can surely expect air crashes to happen each year, while human civilisation still awaits its first asteroid impact. But it helped to raise attention to the impact hazard in wide circles of the USA — from Hollywood directors via insurance business to the very Congress in Washington, D.C. The book by Chapman and Morrison was published in 1989, and in 1992, NASA was instructed by Congress to carry out a large project called *Spaceguard*. This had two parts. The Spaceguard Survey aimed to find 90% of the undiscovered asteroids within 10 years, which would mean about a thousand objects or more. One can say that the goal of Spaceguard Survey was to make us sleep more peacefully at night, because every undiscovered object is a death threat that may be realised at any time. When all the planned discoveries are made, one can likely mark all the objects as harmless in a foreseeable future. The remaining threat will then be much reduced.

The second part aimed to develop defence systems against possible, truly dangerous objects. Within this framework, there was initially an intense lobbying activity by those who wanted to retain and defend the nuclear arms at a time when many strove to have these finally dismantled by the superpowers. I myself belonged to the nuclear arms opponents, and I felt uneasy that the research carried out by me and my colleagues had given inspiration to the old war hawks on the other side of the Atlantic. Their opinion was that the only effective countermeasure was to detonate nuclear devices next to the sinister asteroids and thereby burst them into pieces. In my perception, this was like hiring a mob of gangsters as babysitters when going to the cinema.

The debate about the methods of defence has continued ever since and become more focused on non-nuclear means. But the search project had a quick start and rapidly became efficient thanks to the large resources that were put at NASA's disposal. Consequently, the discovery of Near-Earth Asteroids was speeded up considerably, and the number of known objects rose rapidly throughout the 1990s.

5.4. Dealing with the Impact Hazard

The impact risk has two sides. One is statistical and can be expressed, for instance, by saying that Earth runs a risk of one in a thousand of colliding with a kilometre-sized asteroid during the next thousand years. The other was evidenced through the successful search of the Spaceguard Survey. When tracing the orbits of the newly discovered asteroids forward in time, one indeed found close encounters with Earth. To begin with, these did not involve any risk of collision, but as the number of asteroids grew, the encounters became more numerous, too. Eventually, the situation would arise that such an encounter was found not to be entirely harmless, but that a real impact was within the limits of uncertainty. In this situation, a real impact risk existed, connected to a certain asteroid at a certain occasion.

Like so many others, I should have understood that it was just a question of time, but when it did happen in the autumn of 1997, we were caught off guard. For me, the situation was especially difficult. I had just been elected Assistant General Secretary of the IAU — the International Astronomical Union. My colleague Johannes Andersen was newly elected General Secretary. The practical work of leading the Union during the following three years was shared between us two. Johannes's research dealt with the Milky Way Galaxy and its constituent stars, and if anything important needed to be done concerning asteroids or comets, he put his faith in my assistance.

On December 6, 1997, at the Spacewatch telescope, Jim Scotti discovered a kilometre-sized Near-Earth Asteroid, which received the designation 1997 XF_{11}. This was routinely observed during the following time, and the orbit was determined with increasing accuracy. Johannes and I were not directly informed, since nothing unusual occurred. But on March 11, 1998, everything turned upside down. One issue in the running series of information circulars, which were published by the IAU Telegram Bureau in Cambridge, Massachusetts, contained a sensational note about 1997 XF_{11}. It was stated that the asteroid would pass extremely close to Earth on October 28, 2028, and that an impact was "not entirely out of the question".

This news landed like a bomb on Johannes's desk. The situation became truly precarious since, directly after the issuing of the IAU Circular, it was reported from NASA's centre for orbit determinations at Jet Propulsion Laboratory (JPL) in Pasadena that there was absolutely no risk of an impact. As a consequence, there were also accusations that the statement on the IAU Circular was not only faulty but prone to horrifying people in a totally irresponsible manner. I was of course immediately summoned to help Johannes rectify the situation and save the reputation of the IAU.

All of a sudden, there were two official but discrepant statements about the risk of something that could be an imminent obliteration of human civilisation, which was of course unacceptable. My personal expertise was insufficient, and I did not have access to the computer codes, so I could not discern who was right and who was wrong in case the results did indeed differ. For many years, I had been in contact with the Director of the Telegram Bureau, Brian Marsden, and I simply could not put his competence in doubt. In retrospect, I now see that the differences did not arise from computing errors but from the selection and weighting of the input data, i.e. the observed positions of the asteroid. The conflict was rather of a semantic nature. Marsden had said that the risk was very small but not equal to zero, while the NASA scientists had used the expression that the risk was "essentially non-existent". What's the difference? Shall we emphasise the existence of the risk or its extreme smallness?

The conflict between Marsden and many of the other leading scientists continued for many years. Meanwhile, one asteroid after another was discovered with a non-zero future impact risk. A consequence was one discussion after another about how to disseminate these results so that news consumers of the world would grasp the issue without being misled. We astronomers — like the journalists — gradually learned from these discussions. But to some extent, the IAU efforts for some time felt like Don Quixote's fight with the windmills. A news note from the scientists might look about as follows: "The asteroid so and so has proved to be passing very near Earth on September 27, 2026". The newspaper headlines might then state: "The world will end on 27-09, 2026, astronomers say".

The principle of such risk estimates is the following. The observations of the asteroid are used to determine its orbit at the time when the observations were made. But this orbit is not known exactly. There is an error margin on each of the parameters that describe it. This is ultimately due to the unavoidable errors of the measurements. One can thus present a nominal orbit, but each orbit falling within the error margins might be the real one. Now, imagine that we trace each and every orbit exactly up to a coming close encounter. If one part of the orbits leads to an impact with Earth and the other part leads to a near miss, there is an impact risk, and its size is determined by the size of the impacting part compared to that of the totality of possible orbits.

When new observations are made, in general the error margins of the nominal orbit decrease, and the total amount of possible orbits also decreases. In case all the impacting orbits remain, the impact risk therefore increases. But in the normal, actual case of no impact, all the perilous orbits should eventually vanish, so that the risk is reduced to zero. This is how it looked in several notorious cases, and the astronomers were innocently put in the corner. First, they played doomsday prophets, and then they retracted their horrible prophecies. Were they incompetent, or did they try to raise more money for their research? Sometimes, I actually felt that such suspicions were in the air.

However, this was just a temporary episode that unfolded about 20 years ago. The situation was quickly normalised, as more than one research group developed computer codes that continuously and automatically monitored all the observations of asteroids, made new orbit computations and published the future impact risks on their web pages. Here, the IAU Minor Planet Center played an important role as a gateway for collecting observational data, and in the IAU leadership we also worked hard to make this work adequately. Two independent risk listings are available: one at the Pisa University and one at JPL in Pasadena. No press releases are sent to the media except in exceptional situations, and journalists can at any time get all the information from the web pages. Thus, the interest in the risk estimates has cooled down to a more normal level.

Let us now look at the impact risk judgements from a different perspective. Instead of the asteroid orbit and the parameters characterising it, one may present the asteroid's position in space and its velocity at a given moment of time. If the asteroid is well observed, the position is well determined, and one hence knows that the asteroid (more exactly, its centre of mass) is contained within a small volume around the nominal position. As time passes, this volume moves with the asteroid along the nominal orbit, but in addition, something remarkable occurs. Due to the small uncertainty of the asteroid's velocity, there is also an uncertainty in the period of revolution. This means that the asteroid has a chance to move a little faster or a little slower along the nominal orbit. As a result, the small volume changes shape into a sausage and a string stretching along the orbit. After a few decades, this effect has grown extreme, and the initial volume has become a very long string occupying a larger and larger part of the orbit.

Every point in this string might represent the real asteroid, and we do not know which one it is. We may thus think of each such point as an imaginary or virtual asteroid, and as time passes, these virtual asteroids move along the orbit like a ghost train. If a close encounter occurs with Earth, it may happen that some part of the train impacts our planet. The question now is, is the real asteroid in this wagon?

Let us move back to the Christmas holidays of 2004. I celebrated this in peace and quiet with my wife and my old mother in our Uppsala apartment. On Boxing Day, we saw in the newspaper a note saying that a recently discovered asteroid denoted 2004 MN_4 would pass extremely close to Earth on April 13, 2029. The impact risk was a few percent, and — typically enough — April 13, 2029, is a Friday. I first suspected this to be a canard due to the extremely high risk, by far surpassing all that I had previously come across. But it soon turned out that the figures were correct. On the following day, the risk had reached 1/37, i.e. as high as the chance to win directly by putting the chip on one of the fields of a roulette board. All who have played roulette know that such things happen.

I feared an inferno of excited interviews including numerous misquotes, but another event caused all this to vanish. This was the East Indian tsunami disaster, which occurred on Boxing Day and took the lives of 543 Swedes. This of course caught all the attention, and the impact threat in 2029 was relegated to the small notes. The asteroid was eventually named Apophis, which is the name of the Greek for the Egyptian god of destruction Apep. Its average diameter is only 370 metres, but if it hit a country the size of Estonia or the Netherlands, the whole country would be devastated.

On the third day of Christmas 2004, the predicted ghost train of Apophis's close encounter with Earth was as short as 83,000 km, and the dangerous wagons occupied somewhat more than 2,000 km. But even if the tsunami had not occurred, the issue would have been mute already later the same day. By that time, Apophis had been identified on sky images taken more than 3/4 of a year before the actual discovery. When these observations were included, the ghost train shrank considerably in length, and the whole dangerous part disappeared. This means that, even though Apophis will come extremely close to Earth on April 13, 2029, and possibly reach naked-eye visibility in spite of its smallness, there certainly will be no impact.

Yet, it is still too early to give an all-clear signal. All the virtual asteroids will pass so near Earth that their orbits will be significantly deflected by the Earth's gravity. These deflections are mutually very different, depending on where in the train the asteroid is situated. There are some immensely short segments, where the 2029 deflection will toss Apophis into an orbit, whose period is in exact resonance with that of Earth. As a result, Earth and the asteroid come back to the same place after an integer number of years, so that the collision eventually occurs although somewhat delayed. These minimal segments are usually called key holes, and the risk of the real asteroid being situated there is of course extremely small. For Apophis, the key holes are actually smaller than the asteroid itself, so one issue is where the centre of mass is situated.

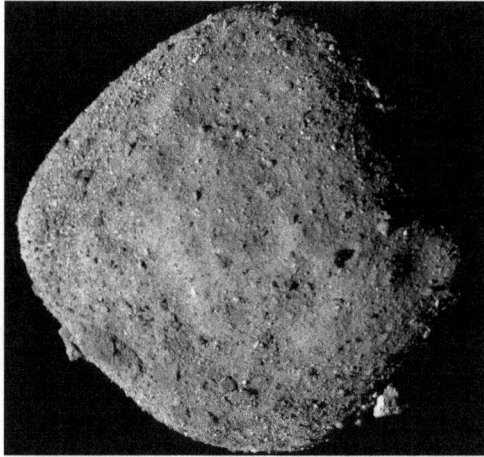

Fig. 5.8. Mosaic image of asteroid Bennu, taken on December 2, 2018, from the OSIRIS-REx spacecraft. Credit: NASA/Goddard/University of Arizona.

Some of these key holes are still relevant. Their total impact risk is currently 1/110,000, and the majority concerns an impact on April 12, 2068.

Apophis is one of the asteroids that have been considered as targets in preliminary studies of techniques for deflection, i.e. techniques that are to lead the asteroid in some way from a risky orbit into one that is harmless. I shall not enter into these issues here — it remains to see if they will ever be used. Quite likely, the most serious current threat is not Apophis but another asteroid named Bennu (see Fig. 5.8). It is predicted that this will pass very close to Earth in a little more than 100 years, more precisely in 2135. On this occasion, there is a risk that Bennu will pass through either of several key holes leading to impact toward the end of that century. The aggregate risk is currently estimated at 1/2,700.

Considering that it takes more than 150 years before the issue becomes urgent, it seems premature to plan for defensive action today. However, Bennu is already being explored by a NASA probe named OSIRIS-REx, as I mentioned in Section 2.7. Its relatively large impact risk is certainly important and often noted in this connection.

5.5. The Great Lunar Bombardment

What can the impact craters on the Moon and the planets teach us about the early history of the solar system? Let us start with the lunar maria. As I have mentioned, these are more than 3 billion years old, but let us now see what this implies. In reality, they originated as wide lava fields that arose when giant impact basins were filled from below by upwelling magma from the Moon's interior. The age of the lava flows — more precisely, the time when they solidified — has been determined by radioactive dating of the basaltic rocks dominating their surfaces. A good example is the Apollo 15 landing in 1971 in Mare Imbrium (sea of rains) (see Fig. 5.9).

This easily recognisable formation on the lunar nearside corresponds to one of the largest impact basins, which is also one of the youngest with an age of about 3.9 billion years. But it took

Fig. 5.9. Image of astronaut Jim Irwin on July 31, 1971, standing by the Lunar Roving Vehicle of the NASA Apollo 15 lunar mission to a landing site in Mare Imbrium. Mt Hadley is seen in the background. Credit: NASA/David Scott.

hundreds of millions of years of volcanic outflows before these were finished and the youngest basalts had solidified. The same story goes for other lunar maria as well. The bombardment causing the impact basins occurred about 4 billion years ago, while the solidified surfaces, on which new craters could be imprinted, are about 3–3.5 billion years old.

Everyone who has studied the Moon with a pair of good binoculars or a small telescope knows that the lunar maria and the highlands look like two distinct worlds. While the maria are essentially even and smooth, the highlands present a chaos of craters. There seem to be no large protected areas. The sharper the images and the higher the magnification, the smaller the size of the discernible craters until the whole surface is filled. Of course, the highlands cannot be older than the Moon itself, and we know that the Moon's age is about 4.5 billion years. Hence, the highlands are no more than 50% older than the maria, so the difference in crater coverage is indeed remarkable.

It has been realised that these observations can be explained if the Moon was bombarded at a rapidly decreasing rate between 3 and 4 billion years ago. In other words, the bombardment was much more intense at the beginning of this period, and toward the end it had settled at the relatively low level that has remained since then. The latter is the current bombardment, but the former was something very different. What was this great bombardment?

This issue has been debated ever since the 1970s, and the debate is still ongoing. As far as the last 4 billion years are concerned, there is hardly anything to argue about, but the situation changes fundamentally when we come to the preceding time, when the Moon was even younger. This is largely due to the fact that the traces of earlier impacts have been erased by the later bombardment. Furthermore, giant impacts of very old age may have left no clear evidence, since the lunar mantle in those days may have been too plastic to preserve the basin-like structures. As a consequence, we are somehow groping in the dark as regards the oldest history of the Moon. Personal opinions may differ, and there is no obvious way to tell who is right.

Two opinions — each one rather extreme — are competing. According to one of these, the wild era 4 billion years ago marked the end of an epoch, which started by a truly extreme bombardment when the Moon was newborn and then became less wild with the passage of time. The projectiles would have been original building blocks of planets — so-called planetesimals — that were left over as the planets grew to completion. According to the other opinion, these planetesimals should have ended much earlier. The initial wild epoch would thus have been followed by a calm period, which lasted until the arrival of a new shower of projectiles 4 billion years ago. What these projectiles were, and how the shower came about, remained enigmatic for a long time.

A decisive event occurred in 2005, when the journal *Nature* published three papers that were intimately connected and were authored by the same team of four scientists. These papers were to revolutionise our understanding of how the solar system planets got arranged into their familiar, orbital configuration. Indeed, more than this, it was actually the first time that we can speak of a real understanding. The theory in question is called the *Nice Model*, and I shall describe it in some detail in Section 6.3. But one of its possible consequences concerned the very bombardment that the Moon experienced 4 billion years ago. The standard term in use for this is the LHB, i.e. the Late Heavy Bombardment. The word "late" refers to the fact that this occurred as late as half a billion years after the formation of the Moon.

According to the Nice Model, the planet migration that shaped our solar system caused an immense number of small bodies — both asteroids and comets — to be flung around by the gravities of the planets. They thus received orbits that crossed the entire solar system. Since this migration was a rather brief affair, the Moon would have been struck by a sudden bombardment without any warning. The authors of the Nice Model believed in the latter of the two mentioned ideas for the lunar bombardment, i.e. the sudden shower 4 billion years ago, and they claimed to have, for the first time, found an explanation. This contributed to a certain lining up of the research community behind an episodic LHB separate from the early planetesimal bombardment. As a consequence, it was then also possible to tell with some confidence when the planet migration took place — something that the Nice Model itself could not inform about.

Since 2005, the Nice Model has evolved and partly changed in several respects. One consequence has been that sometimes asteroids and sometimes comets have been regarded as main contributors to the projectiles. The main site of origin was sometimes the asteroid belt, in particular its inner edge, and sometimes the vast region outside the planetary orbits, where the comet nuclei grew (see Section 6.2). I myself got engaged in this debate and made a brave attempt to show how the water, which flowed on the surface of Mars almost 4 billion years ago, had been brought there at the time of the LHB as a consequence of the Nice Model. But when all the computations had been done and we had the results before us, it turned out that the LHB contribution to the water on Mars, like that on Earth, was only marginal.

This was in line with some results that American scientists had derived from the Apollo lunar samples. These contain lots of melted and re-solidified material from the cratering impacts, and some of this derives from the projectiles. From chemical analysis of the melts, the researchers showed that the abundances of different elements fit very well with a mixture of different kinds of meteorites originating in the asteroid belt. On the other hand, what they did not do was to show that comets would have yielded a worse fit, and this makes me somewhat sceptical.

However, in 2018, a new paper appeared, which introduced important information about the issue. This was authored by a research team led by Alessandro Morbidelli — one of the authors of the Nice Model. The context was as follows. I mentioned earlier that the basalts of the lunar maria once welled up from the lunar mantle and filled the large impact basins. In these basalts, one has found iridium, osmium and other elements (heavy metals), whose presence requires an explanation. This is not trivial, since the Moon is believed to originate from a gigantic collision that struck Earth when its growth was almost completed. The metals in question love the company of iron and are thus called siderophiles, and according to the violent scenario of the Moon's origin, they would tend to concentrate into the iron cores of Earth and the Moon.

Thus, the idea is that the siderophiles in the lunar mantle have been placed there later on in connection with the bombardment by asteroids and comets. This is precisely the mechanism that Alvarez and his team put forward as the explanation to the iridium enrichment in the K/T boundary layer (see Section 5.3). From the measured amounts of siderophiles, one can then deduce the total mass of all the projectiles that have hit the Moon. The expression "all the projectiles" has earlier been interpreted literally — i.e. throughout all times since the Moon was formed. Morbidelli and his team showed that, with this assumption, an extra "shower" of asteroids and comets at the time of the LHB is needed to explain the craters on the Moon. Only the residual planetesimals would not suffice. But they also showed that reality was perhaps quite different.

It is possible that it took more than a 100 million years for the Moon to "calm down" after its violent birth. Initially, the projectiles impacted into a liquid magma ocean on top of a plastic and streaming mantle. Under such conditions, the iron and the siderophiles could be transported down into the iron core of the Moon. The siderophiles in the mantle would hence start to collect only when this time was over. In such a case, there is no need for any late arriving shower of asteroids and comets, and the lunar craters are well explained by the residual planetesimals.

The authors in question did not pass any definitive judgement concerning these two possibilities. But I would rather place my bet on a strongly delayed collection of siderophiles. This would mean that the planet migration of the Nice Model may have occurred at any time during a very long period, and if so, there is good evidence that the relevant time was long before the LHB. As mentioned, together with my colleagues, I had already shown that this migration had very little to do with the origin of the water on Earth and Mars. I will discuss the origin of water in Section 7.1.

Chapter 6

How It All Started: The Birth of Our Solar System

Most astronomers agree that the Universe was born out of the Big Bang. There is also a consensus that the Milky Way Galaxy took shape relatively early — everything in fact happened very fast in the beginning, before the evolution calmed down. As the time eventually arrived for our solar system to be born, the Galaxy was roughly the same as today. This allows a special method to study the birth of the solar system to be used. One simply observes the places in the Galaxy, where stars are being born right now in front of our eyes — and then, preferentially, stars with masses similar to the Sun. One hence assumes that the processes revealed by observing the ongoing star formation also took place 4.567 billion years ago, when the solar system originated.

This seems like an obvious idea, but it was not until recently that it was possible to realise it in earnest. In the 1970s, when I was studying at the Stockholm Observatory at Saltsjöbaden, extensive research was carried out there on the topic of star formation, and the work was partly aimed at adopting new observational techniques, using telescopes on balloons or rockets to catch the infrared radiation from the dusty environments of newborn stars. But this development had just started, and the results obtained in those days were very modest in comparison to what is done today, when the technique stands in full blossom, and new super telescopes receive radiation also in the form of microwaves, which is even more clarifying.

Where, then, do we stand today when it comes to knowledge and understanding? Have we made any progress? Yes, indeed we have, and this is largely due to the fact that our eyes have been opened through new observational techniques. We now talk of transneptunians and exoplanets as though these were trivial things, but in the 1970s we did not even know whether these existed. Thus, in terms of knowledge, we have taken giant leaps. The amount of available facts underlying discussions of the formation of stars and planets has multiplied with every new decade, and thus, these discussions have sharpened. Then, what about the understanding? It is obvious that we understand much more today than we did 40–50 years ago, but yet, we cannot say that we really understand. It is like climbing an unfamiliar mountain. You climb one slope after the other, but you don't know how far the top is.

Still, I will now try to describe how the view looks from our current position — i.e. what we today call our understanding, although it is only preliminary. My attention will be somewhat focused on the role of the small bodies in the formation of our solar system.

6.1. The Spinning Disks

The Universe is full of spinning disks. In the centre of these disks, we sometimes find massive "monsters", which attract the material of the disks by their gravity. The disk survives only thanks to its rotation and the corresponding centrifugal force.

There are many examples of this from our observations of other galaxies, and seen from outside, our Milky Way, too, would appear as a spinning disk, in whose centre a black hole voraciously eats part of its neighbouring material. But the same phenomenon also exists at innumerable places elsewhere in the Galaxy around monsters of lower mass. Our solar system, too, is reminiscent of such a spinning disk, and the central star is our Sun. Now, I'm not going to accuse the poor Sun of being a voracious monster, but once upon a time, it actually was such a thing. Otherwise, it would never have grown.

I am referring to a phenomenon that the theoreticians call accretion disks. If a star such as the Sun is formed out of a gigantic dispersed cloud of gas and dust, this cloud in some way has to collapse and contract under its own gravity. This idea, however, faces a problem since the evolution from a cloud to a star signifies an enormous amount of contraction. Since the cloud initially must have had some rotation — almost no matter how small — this rotation would be speeded up by the contraction, until the gravity was balanced by the centrifugal force, and the collapse ended. Thus, the result is not a star but a much larger, spinning disk.

The solution to this problem is that the material of the disk is subject to an internal interaction, so that the inner parts are decelerated and the outer parts are accelerated. All the material that in this way has been deprived of its motion piles up at the centre, and as time proceeds, there is a permanent, inward transport through the disk. I shall have to pass by the important issue of how this interaction works. The problem is not yet fully solved, even though the first ideas were formulated already 75 years ago. No matter what the mechanism is, it presents itself as a viscosity, i.e. a "tenacity", of the disk's material.

Let us now return from theory to reality. The observations of forming stars have long verified the expectations coming from the theory of accretion disks. As the gas streams inward, toward the centre of the disk, it is heated by friction. The dust is heated, too, and starts to radiate its heat like a stove. As a result, the light from the star is supplemented by a large amount of infrared radiation, which is indeed exhibited by the youngest protostars — so-called T Tauri stars.

From such observations one can also deduce the rate at which the mass of the star is growing. It turns out that this growth is fastest in the beginning and then fades in the course of a few million years. T Tauri stars that are even older often seem to have lost their disks completely. True images of such disks in regions of active star formation have existed for quite some time — largely due to the Hubble Space Telescope. These have become known as "protoplanetary disks" due to the suspicion that they provide just the right environment for the growth of planets. But the old Hubble images are now trounced by high-resolution images using microwaves with a wavelength of about 1 mm. The foremost of these come from the gigantic ALMA Observatory in northern Chile, which is run by the European Southern Observatory (ESO) with participation by the USA and Japan.

The most widely published image features the T Tauri star HL Tauri. It is not a coincidence that both T Tauri and HL Tauri are located in the constellation Taurus, because this is where we see our closest and most prominent star formation region. The image of HL Tauri presents a series of concentric bright rings surrounding the star. One should keep in mind that our eyes cannot register the radiation, which the image serves to illustrate. The light of the image simply shows the intensity of the radiation, and this radiation in turn comes from warm dust grains about a millimetre in size, located in the stellar environment. The bright rings thus indicate the regions where the radiating grains are concentrated. By contrast, the dark regions are remarkably deficient in this dust (see Fig. 6.1).

But the dust is just like the yeast of the ocean waves. The very waves are represented by the gas disk, into which the dust is immersed. Just like the stars emerging from them, these disks are almost entirely made of hydrogen and helium. Other gases are also present but only in very small concentrations. Among these, carbon monoxide (CO) plays an important role as indicator of the gas, since its spectral lines are easily observed by radio telescopes. In this way, the gas disk around HL Tauri was discovered long before the new dust images were produced. However, the structure of the gas disk could not be mapped in detail, and this emerged only recently using the dust as indicator. The fact that the dust is gathered into concentric rings may mean that the gas as well is similarly concentrated.

How can this be? Is there something hidden in the dark zones, which pushes the gas and dust away? Yes, and this may be planets. We understand very well how this could work, though the mechanism seems paradoxical. Due to its gravity, the planet attracts the gas that streams past it on the inside and outside, as seen from the centre of

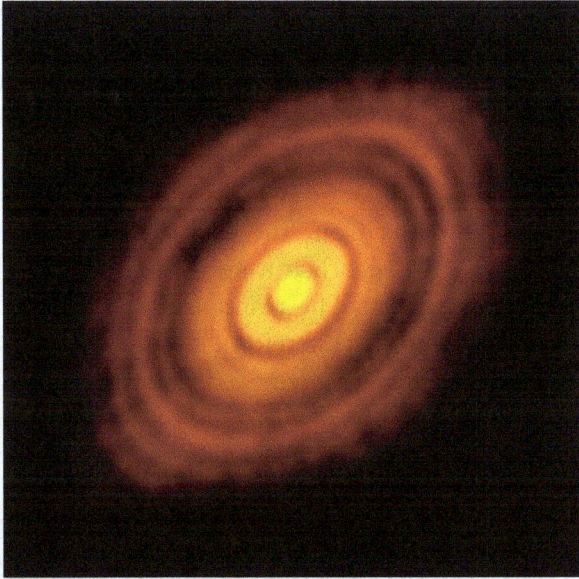

Fig. 6.1. ALMA image of the protoplanetary disk around the T Tauri star HL Tauri. Note the series of bright and dark, concentric rings (see the main text for details). Credit: ALMA (ESO/NAOJ/NRAO). License: Creative Commons Attribution 4.0 International.

the disk. But the result is counterintuitive in that the gas is repelled from the planet's orbit so that an almost empty gap is cleared around it. By the way, there are further paradoxes in celestial mechanics. For instance, take a comet or an asteroid moving on an elliptic orbit around the Sun. Let us imagine that we give it a push forward, as it passes the inner turning point of the orbit. One might think that this acceleration would make it come back earlier for the next passage, but the result is the opposite, since the acceleration enlarges the orbit and increases the revolution period.

The reason for the structure involving bright and dark rings in the disk surrounding HL Tauri is not yet clarified. Different ideas have been proposed to explain the phenomenon. But the planet hypothesis is popular and has inspired much further research. If this hypothesis is indeed valid, the most fascinating consequence would be that we are facing a system of giant planets in the process of formation before our eyes. The protoplanets in question must have reached masses comparable to Neptune in order to push the gas away. But the disk is still very young, and the road to full-scale gas giants like Jupiter and Saturn may still be long. If this were not the case, we probably would have seen the thermal radiation of the giants on the ALMA pictures.

Indeed, when a giant planet is formed, a huge amount of heat is developed. Gas streams from the disk onto the surface of the planet. Upon arrival, this gas has a speed of the order of 50 km/s due to the planetary gravity. The result is a shock, which heats the gas to thousands of degrees. Therefore, a newborn giant planet — or one that is almost fully formed — is glowing hot. Jupiter has had $4\frac{1}{2}$ billion years to cool down but is still radiating away the remainders of the heat that it received from gravity, when it was formed.

In 2018, there were reports from the ESO *Very Large Telescope* — also placed in Chile though south of ALMA — about the discovery of a newborn giant planet located in a disk belonging to the star V 1032 Centauri (also called PDS 70). The mean distance of this planet from its star is more than 20 astronomical units, and the star is smaller than the Sun with a modest brightness. But the planet is still shining brightly thanks to a surface temperature of around 1,000°C. This is the glow of a giant that was born less than 10 million years ago.

Can we now state for sure that giant planets — those of the solar system as well as those around other stars — are formed in spinning gas disks in connection with the birth of the stars? If one chooses to persevere in a critical and sceptical attitude, one can argue that real, visible evidence is still missing. Like the normal practice in courts, one can dismiss the most binding evidence for the lack of absolute proof. Even so, in my opinion, this position is too extreme, at least in the case of the planets. If we cannot yet say that the case is proven, I'm sure that we will soon be there.

This would mean the final confirmation of a general idea about the origin of the solar system, which was nurtured by the natural philosophers of the 18th century (Immanuel Kant and Pierre Simon de Laplace are usually quoted), implying that the planets arose out of a rotating disk of dispersed material surrounding the Sun. This basic idea has enjoyed an almost total hegemony for more than half a century, concerning theoretical models, and the disk in the case of the solar system is usually called the *solar nebula*.

Thus, the solar nebula is now considered to be the birthplace of our planets. But the planets did not suddenly emerge out of nothing. Their origin was a construction project that went on for millions of years. Their predecessors were smaller bodies, and among these we also find the predecessors of comets and asteroids. The next issue to treat is how these, in turn, were formed.

6.2. The Planetesimals

At the beginning, all was gas and dust. This was the case with the cloud from which the solar nebula and the Sun arose. By means of observations of present-day clouds of the same kind, we know that these dust grains were extremely small — their diameters were less than one micron. Yet, these very grains were the first seeds for the growth of the planets. The way from one thousandth of a millimetre to thousands of kilometres indeed seems almost infinite. The growth in question spans 1 billion (10^{12}) times concerning the diameter, and if we estimate how many grains are required to build one planet, the result is 10^{36}, i.e. a number written as a one followed by thirty-six zeros. How is this possible, and how does it work?

The story mostly deals with interactions between the dust and the gas of the solar nebula. To begin with, these hydrodynamic interactions are unchallenged, but as time proceeds, gravity also joins the picture. From the time when solid bodies of several kilometres in diameter have been formed, the evolution has principally been governed by gravitational interactions. These objects are usually called *planetesimals*. Many difficult problems are involved in finding the correct way from grains to planets, but the hardest ones concern the origin of planetesimals.

The start was a delirious dance. We bring forward the magnifying glass and watch a small lump of gas somewhere in the solar nebula. This also contains some of the tiny, original grains. The hydrogen molecules and the helium atoms travel around at high speeds in

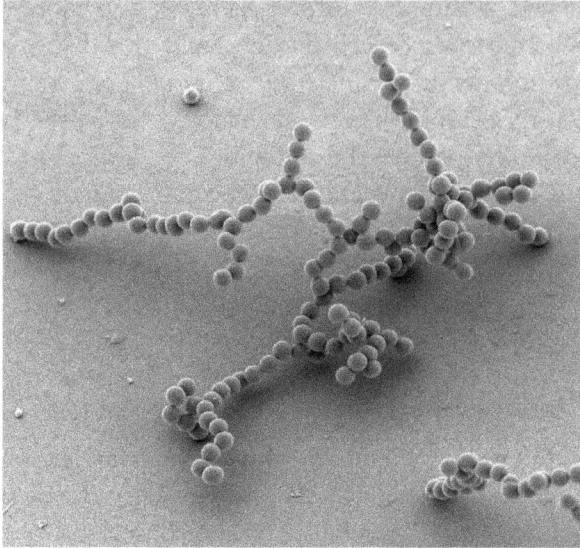

Fig. 6.2. Fractal agglomerate of equal-sized silicate grains with about $1\,\mu$m radius, produced in a laboratory experiment with rarefied air in a levitation drum. Courtesy J. Blum. Adapted with permission from Blum, J., *Space Sci. Rev.* **192**, 265–278 (2000). © 2000, Kluwer Acad. Publ.

their random, thermal motions. They constantly collide with each other, but the bounces are elastic, and the dance goes on without waning. The grains also participate, but even though these are so extremely small — as seen from our perspective — they are huge giants compared to the gas particles. Therefore, their speeds are much lower. But since they indeed move, it sometimes happens that they run into each other and stick together. This is the first step of planet construction (see Fig. 6.2).

Of course, such pairs of grains can also collide with single grains or other pairs so that they stick together. Hence, the process can continue. But a menacing stop sign stands along the road. The clumps of aggregated dust grains appear like very porous dust bunnies, and it is fascinating to think that some of these may recently have been detected by the Rosetta probe close to the nucleus of comet Churyumov–Gerasimenko. I mentioned this in Section 1.6, referring to the extreme porosity of some detected grains. But as the dust bunnies grow in the solar nebula, their increasing mass makes them move slower and slower, until they practically remain at rest. When they have consumed the original grains, they cannot grow any further, because it takes forever to meet another large clump. And yet, these large clumps are far too small to compete with the mentioned millimetre-sized grains in the disk around HL Tauri.

Hence, some other mechanism has to intervene for the growth to proceed. We put the magnifying glass aside and look at a somewhat larger part of the solar nebula. We then discover that the small gas lumps, which we just considered, take part in another dance on a larger scale. They move around in eddies of varying sizes — the smaller ones inside the larger. Physicists refer to this phenomenon as turbulence. When the aircrew signals a risk of turbulence, it means that the plane may enter into a region with such whirling eddies, so that the lift of the plane diminishes, and we experience bumps.

The turbulence of the solar nebula was indispensable for the dust clumps to continue growing. When two eddies meet and the gas is mixed, the clumps may also meet and stick together. This growth eventually introduces a new phenomenon. The dust clumps become so large and massive that the gaseous eddies cannot overcome their inertia any more. Like cars sliding off the road in slippery winter curves, these clumps gradually leave the smallest and fastest eddies behind and only stay in those that are large and slow enough. We may recall the delirious dance and compare the largest clumps with elderly, corpulent and rigid participants on a crowded dance floor. They are all right with a quiet waltz, but in a wild rock'n'roll they tend to be in the way and get pushed around.

In this manner, the growth may in any case proceed, and one might think that it could proceed indefinitely. A hierarchical process, where smaller units merge into larger ones, which in turn merge into even larger ones, would eventually lead to planetesimals. But such is not the case, for new stop signs are found along this road. When a large dust clump grows due to the turbulence of the gas, there is a continuous increase of the relative speed at which it collides with other clumps. Laboratory experiments have shown that there is a size limit, beyond which such clumps do not stick together but bounce upon collisions. This clearly puts an end to the growth. The situation would only become worse for grains that somehow would manage to overcome this bouncing barrier. The continued collisions at even higher speeds would smash the grains, so that they break up into pieces.

We may thus imagine that the spinning disks around nascent stars at present — and the solar nebula as well at the time in question — contain plenty of dust clumps, whose sizes are close to the limit of turbulent growth. As far as we can see, this idea is in accord with the fact that millimetre-sized grains are observed in the disk around HL Tauri. These may have reached the end of the road. But this would only mean that there is another road leading further to planetesimals. In the absence of such a road, comets and asteroids would not exist in our solar system. The existence of Earth would also be a mystery.

According to recent research, the road to planetesimals may be summarised as a three-step process. The first step can be likened to a snowfall. The picture of the solar nebula as a disk, which I offered in this chapter, may be somewhat misleading, since it was not at all flat but extended far from the central plane. The turbulent eddies and their escaping dust clumps were present everywhere. When the clumps had been released from the eddies, they fell toward the central plane under the action of gravity in a way similar to the dust bunnies falling to the floor in our rooms. This gave rise to a new situation, which had important consequences. In the disk at large, the dust only represented one percent or less of the mass, but the snowfall created a thin layer at the central plane of the disk, where the dust was enriched to a high degree.

Inside this layer, the dust clumps moved in orbits around the Sun just like the meteoroids do in today's solar system. But the difference was the solar nebula, i.e. the gas that the clumps had to move in. The gas also moved around the Sun, though somewhat slower, since the gravity of the Sun was partly offset by an outward-directed pressure force. Hence, the clumps had to plough their way through the gas and experienced a significant headwind. The braking effect of this headwind caused them to gradually drift towards the Sun. But this drift was decelerated, in its turn, by the fact that the mass inside the dust-enriched layer was largely constituted by dust. Thus, the gas did not enjoy hegemony but was pushed forward by the dust clumps.

Let us depict the situation. The solar nebula was a gas disk of important thickness, where small dust grains whirled around in the turbulent gas. In the central plane, there was a thin layer of "pebbles", a few millimetres in size, which were slowly drifting sunward. How could planetesimals emerge from this situation? The second step of the mentioned three-step process signifies that the pebbles left their even distribution and piled up in certain favoured regions. There is no mystery about this, because there are reasonable explanations. These always involve a collective interaction between the pebbles and the gas.

When the pebbles get concentrated to special places, they may locally dominate over the gas. If this occurs, there is a phenomenon that may drive the amount of concentration to extreme levels. This is a so-called instability, which may be illustrated by the peloton of a bicycle race. The participants of this peloton feel almost no air resistance, and when the leaders succeed each other, the peloton has a large advantage over individual breakaway cyclists. These are easily caught up, and the peloton grows. In the solar nebula, in addition, new pebbles came drifting from outside. The crowd of pebbles grew faster and faster, until the third step of planetesimal formation took over.

At this stage, the density was so high that the crowd of pebbles collapsed under its own gravity. In the first computer simulations, it turned out that these collapsing crowds often had masses comparable to Ceres — the largest asteroid. This was a little more than 10 years ago. I recall my own surprise. Now, one had found a way for planetesimals to form, and like the other old players in the game, I imagined these as kilometre-sized bodies — about like the usual comet nuclei or small asteroids. But what was formed in the simulations was hundreds of times larger! Time for re-thinking. . .

As to the asteroids, nothing dramatic was needed. We already knew that the asteroid belt is like a crusher (see Section 2.4) and that small asteroids tend to be fragments of larger bodies that have been smashed to pieces by collisions. The story now had to be rewritten such that all asteroids arose as large bodies rather than what the old concept predicted, namely, that they started out small, grew larger by merging at low-velocity collisions and then were crushed, as Jupiter perturbed their orbits and rendered the collisions destructive.

As I shall soon describe, the picture has changed somewhat today. The asteroid belt may have initially been populated by bodies of very different sizes. But the small ones stood no chance of surviving the collisions, and therefore, only the largest ones form the source of the current belt, where almost all the bodies are collision fragments. In the very beginning, however, the belt was peaceful, since the newly formed planetesimals had nearly circular orbits situated in almost the same plane. The collisions that still occurred involved low velocities and thus were not very damaging.

But as the solar nebula disappeared, the stabilising influence on the orbits disappeared as well. Jupiter's perturbations were free to act and soon made the orbits more elongated and mutually inclined. This made the collisional velocities run high, and the crusher started to work. However, another effect was even more important, namely, that the orbits changed so much that many planetesimals came into collision course with the Sun or were flung away from the solar system at close encounters with Jupiter. Hence, the original massive planetesimal belt was only a brief interlude, which was replaced by a much more sparsely populated asteroid belt.

As for myself, I was more interested in the comets. Here, the problems were harder. Due to their very porous structure and significant content of highly volatile substances, it was difficult to imagine that these would once have been situated in the interior of very large bodies and then set free by vast collisions. To my mind, the collapsing pebble crowds did not necessarily have to form just one very large object, but they could instead become fragmented into clusters of smaller objects with sizes down to those of usual comet nuclei. Eventually, with the advent of more efficient computer resources and the possibility to run simulations with better resolution, my idea seems basically confirmed. But this does not mean that comet nuclei have never experienced collisions, as we shall see in the following section.

The new theory for planetesimal formation also had other consequences. Some meteorite researchers have been able to date iron meteorites and achondrites with high precision. These results indicate an unexpectedly old age of these meteorites. In other words, it took a remarkably short time from the origin of Calcium–Aluminium-rich Inclusions (Section 3.4) until the solidification of the melted material in the large parent bodies of the meteorites. Consequently, these parent bodies were formed very quickly, which in turn explains the widespread melting through the decay of the radioisotope Al-26. It would have been more difficult to understand this rapid build-up in case it had to start from the small classical planetesimals. But in the new picture, where very large bodies appear directly, this difficulty goes away.

One may also note the rapid formation of planet Mars. According to recent dating, this planet was almost fully formed already a few million years after the birth of the solar system and the formation of CAI. It is true that Mars is not a planet of the same sort as Earth or Venus, but rather appears to be an embryo, yet the growth of such large embryos in such a short time is still a bit remarkable. If very large building blocks were available very early, this would of course help. A similar argument pertains to the giant planets — in particular Jupiter. As far as we can judge, these were formed in a time similar to Mars, and yet, we deal with more than 300 Earth masses in Jupiter's case. Its solid core of about 10 Earth masses, which attracted the gas, must have grown very rapidly.

From all this, we can see that the first generation of planetesimals was formed when the solar nebula was still young. But there also seems to have been the second generation, which was formed a few million years later and hosted the parent bodies of the chondritic meteorites. In this case, Ceres would likely belong to this delayed generation. Whether the comet nuclei were indeed formed through the collapse of giant pebble swarms or by a later aggregation of leftover pebbles has been debated for several years now. But if the collapse theory is valid, the comets, too, must belong to the delayed generation, because otherwise, they might have melted due to radioactive heating.

6.3. Migrating Planets

I have to say that life as a scientist is exciting. You obtain your PhD, your professorship and maybe membership in Academies, but yet, you never leave the school desk. Practically all your knowledge is perishable, and new truths — just as perishable — keep being produced. Concepts may stay adamantly in one's mind since childhood without being threatened, until one day they are rocked, and one understands that all was a prejudice that has to be abandoned.

For me, the idea about the stability of the solar system played such a role. I was taught that the orbits of the planets are stable. They undergo periodic variations of shape and orientation, but there is no sign that the amplitude of these variations should change significantly — or that the whole orbit would grow or shrink — even during billions of years. The only planet that required special discussion was Pluto, whose orbit penetrates a little inside the orbit of Neptune. But in this case, too, it turns out that Pluto survives for billions of years without coming close to the outermost giant planet. The 2:3 resonance between the orbital periods plays the most important role in guaranteeing this (see Section 2.3).

I saw no escape from this stability. The planetary system seemed like an ideal clockwork, which had ticked uniformly ever since the time when it was born. I thus found it safe to conclude that the planets had been formed in orbits similar to the present ones. But I made a serious oversight. The perfect stability is due to a fundamental assumption that may not be correct, namely, that the solar system has always had the same constituents as now. Specifically, it is assumed that the asteroids and comets have always been as immaterial to the motions of the planets as they are now. I failed to realise that this assumption could have been wrong, when the solar system was young.

Already in the 1970s, a warning bell could be heard against the idea that all the planets had been formed at the same distances, where they are now situated. Actually, the bell may not have been reliable, but it still helped to put the scientists on the right track. The faulty assumption was that the planetesimals were only a few kilometres in size, as explained in the preceding section, and this led to a timing problem for giant planet growth. The constitution of giant planets can be thought of as a core made of rocks and ices and a surrounding mantle of hydrogen and helium. The core is formed first and then captures the mantle from the solar nebula. But forming the core takes time, especially if the planetesimals are small. For the so-called ice giants — Uranus and Neptune — the formation of the cores during the lifetime of the solar nebula presented a formidable problem due to their large distances from the Sun. This caused the planetesimals to be sparsely distributed and to move slowly.

This problem inspired the idea that Uranus and Neptune could have migrated from their real formation sites outward from the Sun, until they reached their current orbits. It was natural to assume that this happened during the infancy of the solar system. My previously mentioned colleagues, Julio Fernández and Wing–Huen Ip (see Section 2.3), embarked upon computer simulations of the interactions between the two ice giants and the sea of planetesimals surrounding their orbits. Thus, they were able to demonstrate the effect that was predicted by the physicist Viktor Safronov in 1969, namely, that these planets can migrate toward or away from the Sun due to these interactions. Each single planetesimal achieves next to nothing, but if the whole population weighs almost as much as a planet, and there is some degree of systematics in their influences, the effect can be large.

The year was 1984. The work that Julio and Wing presented did not cause any immediate revolution in the research community, but the idea was still a ticking bomb. Its first detonation occurred in 1993, when the astronomer Renu Malhotra explained the origin of Pluto's curious orbit. She suggested that Neptune had migrated outward from an origin at less than 19 astronomical units from the Sun. Pluto would originally have had an almost circular orbit at a distance of

25 astronomical units. When Neptune reached 19 astronomical units, Pluto was captured into its present resonance, which means that it completes two revolutions around the Sun in the same time as Neptune completes three. When Neptune continued to migrate out to its present distance of 30 astronomical units, Pluto was forced to migrate out to 40, and during this migration, its orbital eccentricity also increased to its current, high value. We may note that many other celestial bodies — the so-called plutinos — share the general properties of Pluto's orbit. Malhotra's explanation is thought to apply to all of these.

Malhotra may have been inspired to the idea of a migrating Neptune by the work of Fernández and Ip. Her own work is solid and has not been questioned. But she had no reason to discuss the question how Neptune was originally placed in the midst of the planetesimal sea — something that was to be explored about 10 years later. Before then, new and revolutionising discoveries would be made.

In 1995, for the first time, it was clear that a planet had been discovered in an orbit around another star with properties reminding of the Sun. The work was performed at the French Haute Provence Observatory by Michel Mayor and his PhD student Didier Queloz. These two scientists were in fact awarded the 2019 Nobel Prize in Physics for this discovery. Two of the planet's properties — the mass and the revolution period — became subjects of discussion. The mass appeared comparable to that of Jupiter, which was no surprise, but the orbital period was only 4 days, while that of Jupiter is 12 years. This means that the orbit of the planet is placed very near the surface of the star, which was completely unexpected. As yet, we do not know much about the physical or chemical nature of the planet, but it has been officially named Dimidium.

The expected Jupiter-like mass of Dimidium does not mean that its composition is known. However, it is considered most likely to be a gas giant, much like Jupiter, which consists mostly of hydrogen and helium. The big issue is how such a gas giant can reach so close to its parent star. There is really only one reasonable explanation, namely, that the planet was born much further away and then migrated inward. But this migration had nothing to do with planetesimals — the mechanism was very different. The driving factor was the gas in the accretion disk, where the planet was formed.

Giant planets must be formed in gas disks in order to receive their contents of hydrogen and helium. This was clear before the discovery of Dimidium, and theoretical work had already been done on understanding how the gases are captured and how the planet interacts with the gas disk via its gravity. It had been realised that this interaction could cause the planet to migrate according to different patterns. One such pattern — called Type II migration — was interesting in the case of Dimidium. The reason for this migration is the same effect as is discussed as an explanation to the dark gaps between the bright rings in the HL Tauri disk.

The planet thus clears an almost empty gap around its orbit. But its mass is very small compared to the mass of the entire disk, and it therefore sits in its gap as an encaged little kid. If the disk is in the process of streaming into the star, it continues doing so without any notice of the planet. The gap follows the inward streaming of the gas, and the planet must follow, too. This is Type II migration.

After Dimidium, many other so-called exoplanets have been discovered, whose orbits around other stars remind of Dimidium and whose masses are comparable to that of Jupiter. It is true that such planets are particularly easy to discover, but it is nonetheless clear that they represent a phenomenon that is not very rare. It thus appears that the Type II migration, which should affect all giant planets, sometimes but only sometimes leads to the extreme Dimidium-type orbits. These planets are called *hot jupiters*, where "jupiters" stands for the large mass and "hot" stands for the closeness to the star, which is expected to heat the planetary atmospheres to more than a thousand degrees.

Consequently, it is clear that our solar system is not at all unique when it comes to hosting giant planets but, yet, is not like all the others. Our Jupiter too should have undergone Type II migration, but this must have been stopped so that the planet stayed at a considerable distance from the Sun. It might be thanks to this fact that Earth was able to grow in peace and quiet far from any disturbing giant planet. Of course, the question is what decided if the solar system would take its actual shape or derail into something completely different, where the origin and evolution of life might have been prohibited? Here, we approach one of the corner stones of our current understanding of our solar system — the so-called Nice Model.

I have always enjoyed myself at the Nice Observatory. This may sound like a truism, but it is not only the charm of the city of Nice, the beaches and the mountain villages. The observatory has also hosted eminent scientists, from whom I have learned a lot. In the beginning of the 2000s, a research group worked there with a congenial idea. I knew two of its members since earlier. Alessandro Morbidelli had attended some lectures of mine as a PhD student in the 1980s and 1990s, and now he had outgrown us all. Hal Levison was a congenial chap with an outstanding research capacity, whom I liked to take to my favourite restaurants. The two younger ones, Kleomenis Tsiganis and Rodney Gomes, were no acquaintances of mine, but I often saw them on the bus and at the observatory (see Fig. 6.3).

In 2005, they presented the results of their work, which I described in Section 5.5. The scenario for the evolution of the solar system, which these papers described, has ever since been referred to as the Nice Model. The gist is that the origin of the four giant planets placed them into a rather compact configuration, i.e. in a dangerous vicinity of each other's orbits. By their gravity, they quickly removed all residual planetesimals from the region around their orbits. When this happened, the solar nebula still existed. But when the gas disappeared, this system of giant planets was left to its fate, i.e. to the effects of their mutual gravitational interactions. Had this been the full story, the same tight system might still exist, but something happened, which made the system break down and be transformed into what we see today.

The cause of the breakdown and transformation was a huge population of planetesimals, which formed a disk starting somewhat outside the planetary orbits. No giants had formed in that region, and the growth of large bodies had been stalled at an early stage. Before 2005, the conviction remained that the planetesimals were formed

Fig. 6.3. The original authors of the Nice Model. From left to right: Kleomenis Tsiganis, Hal Levison, Alessandro Morbidelli and Rodney Gomes. Published with permission from K. Tsiganis.

very small and then gradually merged into larger bodies. Because of the large distance from the Sun, these were expected to be largely made of ice. It was thus a question of preliminary stages of comets. Among the most advanced construction projects, we find Pluto and Eris, many more of similar sizes, and perhaps even larger bodies. For an old comet scientist, the Nice Model was like a blessing — it was thus comets that drove the entire solar system into its right path (see Fig. 6.4).

I shall not describe the detailed sequence of events when the comets brought chaos into the planetary system, since these details have not been established. On the other hand, it is clear that at least one giant planet was kicked around like a football by Jupiter and Saturn at close encounters, and that at least Neptune was brought into the outer comet belt in connection with this. Thereby, the whole belt was dispersed at the same time as Neptune's orbit migrated out to its present distance and got practically circularised.

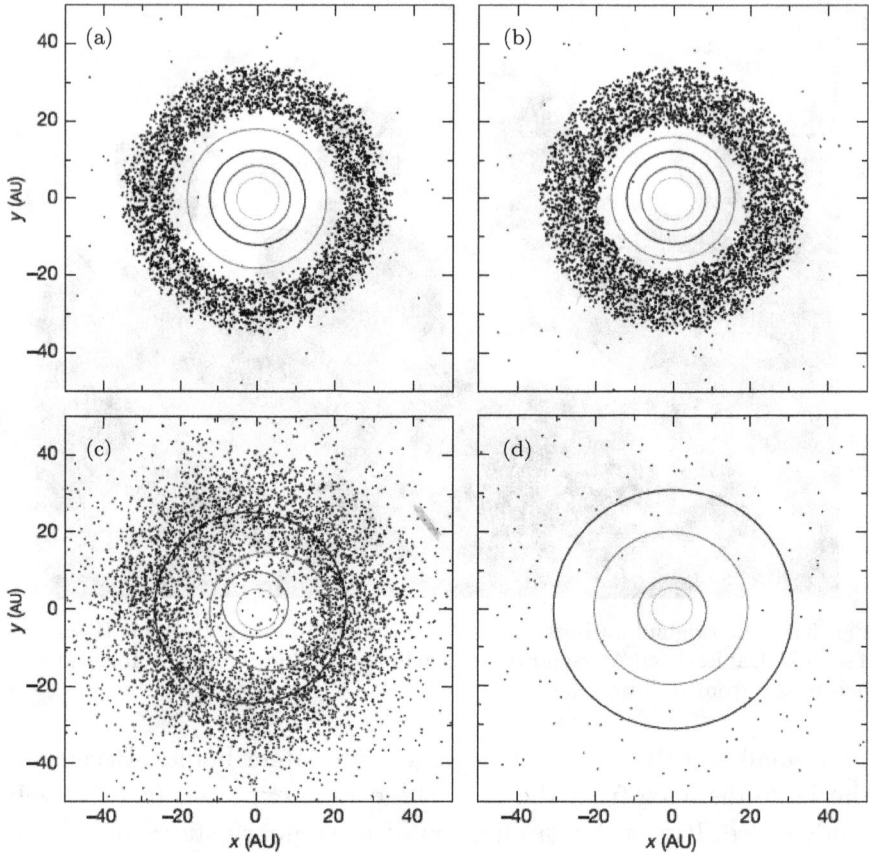

Fig. 6.4. Planetary orbits and disk particle positions in a reference Nice Model simulation of the original work, published in 2005. The picture shows four snapshots at: (a) an early epoch; (b) a time shortly before the onset of the instability; (c) a time during the chaotic period; (d) an epoch representing the current conditions. Adapted with permission from MacMillan Publishers Ltd: Gomes, R. *et al.*, *Nature* **435**, 466–469 (2005). © 2005 Nature Publishing Group.

This may provide the proper background to Malhotra's idea about the evolution of Pluto's orbit.

As already mentioned, the impact of the Nice Model had several reasons. One very important feat was to find, for the first time, a logical explanation to the orbital properties of the solar system planets. This was the foremost reason, but a few others are also quite important. One of these dealt with the Jupiter trojans, i.e. the asteroids that more or less share the orbit of Jupiter and dwell in two regions ahead of and behind Jupiter at about 60° distance (see Section 2.2).

The issue is whether the trojans are primordial or not. When Jupiter was formed, it scattered the primitive planetesimals around by its gravity. Thereby, some of these could have been captured into trojan orbits, and they would be primordial having followed the planet since its origin. But identifying them with the current trojans has a serious problem, namely, the difficulty to explain the high inclinations of many trojan orbits. Here, the Nice Model came to the rescue. The simulations showed that all the old trojans disappeared when the giant planet orbits became unstable. The chaos that struck the planets, struck also the trojans, which escaped in many directions.

On the other hand, new trojans were captured from the huge amount of icy planetesimals coming from the exterior comet belt. The simulated result of this proved to fit very well with the observed number of trojans and the inclinations of the orbits. We may thus consider that the trojans are not asteroids at all but captured comets. For billions of years, their surfaces have been so clogged by dust and dark organics that no activity remains, but at depth, we can forecast that they contain large amounts of ice. It remains to be checked if this is correct.

The Nice Model has risen to an unusually high status and has been regarded as somewhat of a holy grail. Now, we have learned that even those may fall, so the question is if there will come a day when, even for the Nice Model, all proves to be a mistake? I cannot know, but I do not think so. There are too many predictions from the Nice Model, which fit accurately to reality and which are hard to copy by any other theory. But there is also one conclusion that may have been unwarranted and faulty. Typically, it probably was this one that brought most glory to the Nice Model.

The issue is the time when the planetary system broke down. This cannot be predicted by the Nice Model — in principle, it can be anything from 10 million to 1 billion years after the vanishing of the solar nebula, depending on details that we do not know. But, as I described in Section 5.5, it was easy to adopt the idea of an abrupt and temporary bombardment of the Moon 4 billion years ago. From something nearly inexplicable, this would become a natural consequence of the Nice Model, if only the breakdown happened to linger for 400 million years. This would be a beautiful outcome and became very popular, but as I indicated, it could have been a premature conclusion.

How much time it really took is of importance for the evolution of the comets. The comet nuclei originate from icy planetesimals that were formed in the solar nebula throughout the region where the temperature was low enough for ice to survive. This region outside the so-called snow line was very extensive, and the local temperature varied within wide limits. One may therefore imagine different kinds of comets to exist with somewhat different compositions depending on how far from the Sun they were formed. The issue of whether the real comets exhibit any such differences has been debated for many years, but as yet, the results are modest.

The comets that were formed closest to the Sun ought to have had a lot to do with the giant planets. Some of these were engulfed by the planets, while others were thrown into new orbits at close encounters. This happened during the time, when the solar nebula still existed. This fact caused a sorting of the planetesimals according to size, since they reacted differently to the gas drag of the nebula. For the very large ones, this meant nothing, so these were free to stroll all over the solar system and be ejected into the Galaxy. But those measuring only a few kilometres should have piled up in two regions — one just inside and one just outside the zone that had been swept clean by the giant planets. I shall return to the first category in Section 7.1.

The rest of the icy planetesimals had their abode in the wide zone beyond the orbits of the giant planets. This is where we find the origin of essentially all the comets that we observe today. But this implies an interesting problem. The zone was certainly wide, but the planetesimals were so incredibly numerous that some congestion still arose. Collisions may have occurred, and these threatened to smash the planetesimals and grind them into gravel. This risk of course grows more serious if the congestion had to last for 400 million years before the planetary system "decided" to break down.

I brought up this issue with Alessandro Morbidelli during a stay at the Nice Observatory, and together we investigated the risks that the future comets would run. This was at the time when the close-up images of comet Churyumov–Gerasimenko started to arrive, so we looked in particular at planetesimals of that very size. Our results demonstrated something which not many would have imagined. The chance for such planetesimals to survive for 400 million years was very slight. From this, we could draw an important conclusion. The nucleus of the Rosetta comet would not be a primordial planetesimal but instead had to be a fragment of a larger planetesimal that had been smashed up at a collision. Keeping in mind the curious shape of this nucleus, there also followed the idea that we see two fragments that stuck to each other inside the plume of material from the point of impact.

It would indeed be awkward if the basic aim of the Rosetta mission, namely, to explore an unmodified witness from the birth of the solar system, would have been missed by flying to a comet nucleus that had arisen from a violent collision. But is this reasonable to imagine? Most of my Rosetta colleagues, and in particular those of the imaging team, claimed that this cannot be so. The comet nuclei show signs of a far-reaching pristineness that appears to exclude such an origin. Could collision fragments retain the very high porosity? How could extremely volatile substances like nitrogen and carbon monoxide avoid being vaporised and disappearing into space? These questions have been widely discussed. Some scientists claim that the comet nuclei may very well have experienced collisions without losing their virgin charm, while others remain strongly sceptical to this idea.

But the basis for the discussion is torn away in case the breakdown in the system of giant planets arrives very early. The entire huge population of planetesimals would then have been dispersed after such a short time that very few collisions could take place, and the majority of the comet nuclei would thus be just as primordial as we like to think. As to myself, I actually believe that such was the case. But even so, this does not mean that the Rosetta comet has to be undamaged. In the course of its billions of years in the scattered disk, it may have experienced impacts that influenced its shape and spin according to some recent research.

6.4.   The Outskirts of the Solar System

When we talk about the planetary system, we mean the system of the eight known planets, which ends with Neptune's orbit at 30 astronomical units from the Sun. If we also count the Kuiper Belt, we reach out to about 50 astronomical units. If, in addition, we also include the scattered disk, the maximum distances are much larger, but the inner turning points of the orbits cannot be situated too far outside Neptune's orbit, which means in particular that the objects remain under Neptune's gravitational control.

If we consider comets, some of them have orbits that reach extremely far away. These are the new Oort cloud comets and some of the revenants as well (see Section 4.1). However, being observable comets, the inner turning points of their orbits have to be contained within the realms of the planetary system. Let us now turn our attention to another type of solar system objects, whose orbits share an important characteristic, namely, their location entirely outside the influence of the planets. As a rough guideline, the minimum distances from the Sun (the astronomers call these the *perihelion distances*) are about 60 astronomical units or larger. I will refer to this type of objects as defining the outskirts of the solar system.

I have mentioned two categories of objects belonging to this outskirts population in preceding chapters. One is the Oort cloud, i.e. the source of long-period comets identified by Jan Oort in 1950. The other contains the sednoids (Section 2.3) including their namesake, minor planet Sedna. These two categories basically differ in orbital periods, reflecting the sizes of the orbits. For the sednoids, a typical value of the mean distance from the Sun is 500 astronomical units, and for the Oort cloud comets, the values range from a few thousand astronomical units in the inner core to tens of thousands in the outer halo, from which the new comets derive.

As I mentioned in Section 2.3, three sednoids have been discovered so far, but in the case of the Oort cloud, no actual discovery has yet been made. However, we see a constant stream of comets from the Oort cloud, because the large distances make the orbits unstable against the perturbations by passing stars and the Galactic tidal effect. The orbits of the sednoids are too close to the Sun for such effects to have any influence. These objects are thus doomed to stay forever far beyond Neptune's orbit. The total mass of the sednoids has been estimated to be similar to that of the Kuiper Belt, i.e. no more than one percent of Earth's mass, while the mass of the Oort cloud is thought to be similar to or to exceed the mass of Earth.

There is one open problem of large significance which concerns the sednoids. A strange coincidence has been noted regarding the orbits of all three sednoids. Their orbits cross the plane where the giant planets move, along lines that nearly coincide with the lines between their inner and outer turning points. The same property is also shared by the scattered disk objects with the longest known orbital periods. On the other hand, no such coincidence should exist unless there is a reason for it. A speculation arose that the reason could be a planet whose gravity perturbs the orbits of all these distant objects.

At Caltech in Pasadena, the previously mentioned Mike Brown and Konstantin Batygin investigated this further, and they presented results that have attracted lots of attention. These confirmed the idea that the orbits of the distant objects are "shepherded" by a large planet, which itself moves on a similar orbit. The reliability of this prediction is under debate, and it would certainly be unwise to take the planet's existence for granted. As yet, the term generally used to describe it is *Planet Nine*. In fact, the solar system used to have nine planets until 2006, when Pluto lost its planetary status. Currently, many again tend to count nine planets, though the ninth one remains to be discovered.

Today's situation presents some similarities to the one that prevailed before the discovery of Neptune in 1846. John Couch Adams and Urbain Le Verrier had independently predicted the planet and its position in the sky, and Neptune was thereafter promptly discovered at the Berlin Observatory by Johann Gottfried Galle. This time, according to current estimates, we may have to wait about 10 years before the issue of Planet Nine can be settled by an actual discovery or by failed searches, because the planet may be situated at a very large distance. Its mass is predicted to be 5–10 Earth masses, so it would likely be counted along with the giants.

We now come to the most fundamental problem of all the outskirts populations. It concerns the Oort cloud, the sednoids, and Planet Nine as well, if it exists. This comes from the fact that these objects cannot have been formed in the outskirts. There is no way even to imagine such a birth, so we have to assume that they had a normal mode of origin in the planetesimal populations within or just beyond the planetary system. As a consequence, they must have been transported outward in the sense that their perihelion distances were increased way beyond the values that planetary perturbations could achieve. The question is, which agent was responsible for this increase and, thus, for decoupling the objects from the gravitational influence of the planetary system?

There is a property of planetary dynamics that may easily slip one's attention but is very useful to keep in mind. As long as only gravitational forces are involved, for every evolution that may occur, the inverse is also possible. The direction of the flow of time does not matter. In the present case, let us consider the Oort cloud. In Section 4.1, I described the current ideas about how comets of the Oort cloud are led astray so that they enter into orbits that make them observable. In principle, the inverse process can work, too, and it may have done so in the old days, when the solar system was young. Thus, there is a chance that comets were scattered by the giant planets into orbits reaching far away, and these were then extracted into the Oort cloud by a combination of the Galactic tide and passing stars.

This idea has been pursued ever since the 1980s. Through numerous computer simulations, it has been established that an Oort cloud should indeed have been formed with more or less the properties indicated by our observations. But it has not been possible to demonstrate that the efficiency of the process is large enough to explain the expected mass of the Oort cloud, and this is still a lingering problem. The current state of the art is represented by a paper published in 2013 by Ramon Brasser together with Alessandro Morbidelli. They looked in particular at the creation of an Oort cloud as a consequence of the Nice Model.

The time was supposed to be about 4 billion years ago in connection with the Late Heavy Bombardment (Section 5.5). The instability of the giant planet orbits led to the formation of a primordial scattered disk — the parent of the present scattered disk, which is just a bleak shadow. Brasser and Morbidelli simulated the influence on this disk by the Galactic tide and passing stars as well as the continued eroding effects of Neptune ejecting objects from the solar system. They claimed to find a reasonable fit to the masses currently estimated for the scattered disk and the Oort cloud, thereby solving the efficiency problem, but an ultimate judgement on this solution has not yet been made.

Let us now turn to the sednoids. Here, the situation is quite different. These objects are currently prevented from entering into the inner solar system, and yet, there must have been a time when they were allowed to do the opposite, i.e. to get extracted. This means that there has been a fundamental change in the environment of the solar system after that time. It is clear that the Galactic tide or the usual passing stars of our time are too weak to have any influence on the relatively small orbits of sednoids.

The solution to the sednoid problem has a familiar name: *stellar clusters*. Let me illustrate these by introducing the Pleiades, i.e. the Seven Sisters in the constellation of Taurus (see Fig. 6.5). There is romanticism in the beautiful sky of cold winter evenings in the northern hemisphere, and I think that the Pleiades contribute to this. The Pleiades in fact constitute a stellar cluster. By this we mean a local concentration of stars, mostly somewhere in the Galaxy. We know for sure that such concentrations do not arise by coincidence, but the stars were born together at the same place and the same time. The age of the Pleiades is about 100 million years. Using the naked eye, at best, one can usually see seven stars, but the total number is estimated to be between 500 and 1,000.

Fig. 6.5. The Pleiades, a stellar cluster in the constellation of Taurus at a distance of 400 light years, imaged by the Hubble Space Telescope. Credit: NASA, ESA, AURA/Caltech, Palomar Observatory.

There are many stellar clusters, and most amateur astronomers can probably mention several. Yet, in fact, the clusters contribute only a very small fraction of all the thousands of stars that are visible to the naked eye. From this it is easy to see that a large majority of the stars in the Galaxy are either single like the Sun or members of tiny systems like binaries, triplets or the like. But the question remains: were the stars born like this, or have they escaped from the clusters in which they were born? The answer depends on how long the clusters survive.

There appears to be a statistical correlation between the lifetime of stellar clusters and their sizes. This is natural, since all clusters become decimated as time passes, since stars get kicked out and escape. Clusters that are born with tens of thousands of members are able to survive for billions of years. Those which are of about the same size as the Pleiades and start out with several thousand members in general survive for a few hundreds of millions of years. Much smaller clusters live only for much shorter times, and the very smallest become dispersed almost directly after the formation of the stars. From studies of the youngest clusters, which we find in star formation regions, we have learned that the smallest clusters are the most common. The larger the clusters are, the more rarely they appear. It also turns out that the large majority of stars are born in clusters. Thus, a typical feature of the lives of stars is that they are born in large families but die as loners.

Every stellar cluster — whatever its size — seems to be born in a fog of gas and dust that is so thick that light essentially cannot penetrate. The very youngest clusters are therefore referred to as "embedded", and the stars are invisible to our eyes. But they still emerge, for instance, on X-ray images, so we do have some information about them (see Fig. 6.6). Moreover, the fog is dispersed and the gas is blown away already after a few million years so that the cluster can be observed by usual telescopes.

Ramon Brasser and his American and Canadian colleagues have carried out a number of investigations where they have established that embedded clusters may possibly provide the necessary mechanism to extract the sednoids. In any case, this would not work if the Sun had been formed in isolation. Thus, we have to accept the concept of the Sun's *birth cluster*, and it only remains to try to find constraints on its size and other properties.

Let me briefly describe my own understanding of the physical processes involved. According to the observations, the very youngest clusters seem to be out of equilibrium. This means that the stars move around with velocities that are too small to offset the gravitational attraction that tends to pull the stars together. The clusters therefore contract, and one can expect that the densities increase especially in the central regions. This has two consequences. The first is that the gravitational force from the entire cluster varies rapidly with distance from the centre, and this causes strong tidal effects in the environments of individual cluster stars. For instance, planetesimals that are scattered into large orbits by the giant planets — if we imagine the Sun to be such a cluster star — would

Fig. 6.6. Image of part of the Orion nebula acquired by the Chandra X-ray Observatory, featuring among others the very young Trapezium cluster. Credit: NASA/CXC/Penn State/E. Feigelson & K. Getman *et al.*

have their perihelion distances increased to the point where they become sednoids or Oort cloud members.

The second consequence is that, as the cluster contracts and the stars come closer to each other, the frequency of close encounters increases. These may play the same role as the cluster tide and thus contribute to populating the sednoid or Oort cloud regions, but they also may assist in the storage of bodies. This is because the tide works cyclically like a pump — the perihelion distances increase and decrease periodically, but an intervening stellar encounter can break this pattern and save the objects permanently in the outskirts of the solar system.

The fact that the gas is blown out of the cluster within a few million years is of fundamental importance for the cluster, because it deprives the cluster of a large part of the gravitational attraction that binds it together. In fact, the initial contraction always leads to a rebound so that the cluster starts expanding, and when the gas goes away, there is not enough gravity to stop this expansion. Hence, the normal behaviour is that the cluster flies apart and stops to exist.

From the work of Brasser and colleagues, it appears that the violent dynamics of the embedded solar birth cluster may have produced a sednoid population with a mass of one percent of Earth's mass, and this seems relevant to explain the observations of these objects. However, the Oort cloud objects that are also produced do not suffice to account for the estimated current mass of the Oort cloud and, in any case, the extracted objects would be much larger than the new comets that we observe. Hence, the explanation for the Oort cloud may still be the one that Brasser and Morbidelli proposed.

But the timing of the events now becomes an interesting issue. If the Nice Model instability did not occur at the time of the Late Heavy Bombardment, as Brasser and Morbidelli assumed, but already a few tens of millions of years after the birth of the solar system, the extraction of Oort cloud comets may have taken place in a completely different environment. If the solar birth cluster was not as small as Brasser and his colleagues assumed, it could have been more long-lived and remained in existence at the time of the Nice Model instability. The Oort cloud could then have been formed as a result of the cluster tide, and the process may be expected to have been much more efficient than previously estimated. I do not know what the truth is, but further investigations may lead to an answer.

Finally, the purported Planet Nine may also provide useful insights, in case it exists. On the basis of current results, we would say that it is a failed giant planet that was ejected by Saturn and then extracted during the embedded stage of the birth cluster. But this cannot be true! The number of planetesimals has been so immense that the extraction probability of the cluster tide was large enough to explain the sednoids, but it is far from sufficient to extract one giant planet out of a few. Thus, the hypothetical discovery of Planet Nine would speak in favour of a more long-lived birth cluster, where both the Oort cloud and lots of sednoids including Planet Nine could become extracted. Once more, only the future will tell, but this future is exciting indeed.

Chapter 7

Water, Life and Death:
Existential Aspects of Comets

Does the sky have an existential significance? Yes, no doubt. Heaven is mentioned in the first line of The Lord's Prayer. In the Christian view, the power of life and death through God the Father resides in heaven. Elsewhere, for insight into their fate, people often turn to augury. Sibyls augur from coffee grounds and palms, while astrologers tell the future from the positions of celestial objects.

I remember a summer's day in 2004, when my wife and I were travelling by car, northbound through the Czech Republic. At that time, the Alps and Central Europe were struck by one of the foulest weathers in living memory. That same morning, we had left České Budějovice half an hour before the deluge would have taken our car. A little before Prague, on an elevated road section, we were watching the beautiful city. But it was hunching beneath the most gruesome sky I have ever seen. Thunderheads were towering as far away as one could see, and in the midst of the day, buildings and towers stood as if in evening darkness. I cannot express in words what touched my mind the most, but there was something apocalyptic about the whole scenery.

This was an extreme situation, but our state of mind is always influenced by the appearance of the sky. Dark thunderheads make us scared and gloomy, while nice weather inspires hope and trust. In a concrete sense, the sky can bring both life and death. It waters our fields by rainfall, and it can send deadly lightnings and whirlwinds. Geophysicists may account for all such things, but I instead focus on the astronomical side of the issue. Thereby, I turn my attention to astrobiology, i.e. the science of life's place in the Universe. In particular, part of this is our knowledge about the origin and evolution of life on Earth, and how these are governed by external influences having to do with the solar system and the Galaxy.

7.1. The Delivery of Water

All the water on Earth amounts to about one per mille by mass. Thus, the water is far from being the main component, but the issue of its origin is still very interesting — not only due to its biological importance. All the local planetesimals in the vicinity of the Earth's orbit were formed at such a high temperature that they ought to have been next to bone-dry. The planetary embryos that grew from these in the course of a few million years did catch some water vapour along with the gas from the solar nebula, but this primary atmosphere was quickly lost. We have evidence from observations of protostars and newborn stars that such was the case. In addition, we know that the noble gases in the terrestrial and Martian atmospheres appear in proportions that are very different from those of the solar nebula.

Therefore, the present atmospheres and hydrospheres must have a different origin. This origin was, on one hand, a bombardment of the planetary embryos by leftover planetesimals, and on the other hand, giant collisions between different embryos. The time when these events took place started with the disappearance of the solar nebula. It was a rain of planetesimals, which started out very intensely and then gradually waned, as the number of planetesimals decreased. The timescale of this waning was as long as several 100 million years. For Earth, it was mainly a question of building the planet by means of collisions between embryos. A few tens of embryos were required for this, and the process took about 100 million years. Mars, on the other hand, was essentially left as a single leftover embryo.

While the native embryos — i.e. those that did not arrive as migrants from regions further away from the Sun — were dry to begin with, they soon swept up water from wet planetesimals. In the case of Earth, some migrant, wet embryos also contributed water at giant collisions. With a high probability, this is the way that both Mars and Earth acquired their water. The time in question likely ended more or less when the construction of Earth was finished. Likely, the subsequent, continued rain of planetesimals only contributed small amounts of water.

How did the water get into these planetesimals and embryos? The following answers are preliminary, since research is ongoing, but they represent a common opinion. One clue comes from meteorites (see Section 3.3). Enstatite chondrites are extremely dry and, yet, these are considered the best analogue of Earth's main construction material. Water is also very sparse in ordinary chondrites, and thus, a fraction of the planetesimals and embryos with this composition cannot have made any important contribution to the watering of our planet. The interest is rather focused on carbonaceous chondrites, which often exhibit an important water content.

The parent bodies of these meteorites are situated in the outer part of the asteroid belt and, possibly, even further out. Roughly, it is a question of the regions beyond 2.5 astronomical units from the Sun. It is likely that these bodies stem from the planetesimals that originally populated the same region. Hence, we find there the source of the water in Earth and Mars.

The next question is why the water was present there and from where it primarily originated. This is closely connected with the so-called *snow line*, i.e. a critical distance from the Sun, which separated the solar nebula into two zones. In the inner zone, any ice would sublimate into water vapour, and in the outer zone, any vapour would condense into ice. Bodies containing ice must either have been formed beyond the snow line or, otherwise, have collided with icy bodies from the outer zone. From this perspective, there are two different ideas, both of which might explain why water-rich planetesimals were present only beyond a distance of 2.5 astronomical units.

One idea is built on the drifting pebbles that I wrote about in Section 6.2. Far from the Sun, such pebbles must have been partly icy. Some of these ended up in giant swarms, which collapsed into icy planetesimals. But others continued drifting toward the Sun and thus approached the snow line. In case they reached this place, they were rapidly consumed by evaporation. But before this, they also had many opportunities to collide with planetesimals along the way. Now, the question arises, where was the snow line situated?

The observations of disks surrounding T Tauri stars of solar type indicate something interesting. During the quiet phase that follows, when the gas flow into the star is considerably reduced and the stellar mass approaches its final value, very low temperatures prevail near the central plane. The starlight is blocked, and the frictional heating is weak. In such cases, the snow line may fall as close to the star as the Earth's orbit does in relation to the Sun. Such was likely the case in the solar system, too, way back in time. One might therefore think that, at some stage, the drifting pebbles were able to deposit their ice onto the planetesimals near Earth's orbit. But, as mentioned, meteorites tell a different story.

Alessandro Morbidelli and his co-workers have proposed a possible reason for this. While the gas flow to the Sun waned and the solar nebula cooled off, the snow line moved towards the Sun. On one occasion, its distance was 2.5 astronomical units. At this point, a decisive event occurred. The growing planet Jupiter reached the critical mass, at which point it cleared a gap in the gas disk (see Section 6.3). It was then located much further out. As the gap was opened up, neither the gas nor the pebbles could stream past the gap. This put a stop to drifting icy pebbles in the vicinity of the snow line, as this moved closer to the Sun.

There was probably nothing beyond coincidence which forced the growth of Jupiter to occur at exactly the necessary pace, so we can regard it as a coincidence that the limit of water-rich planetesimals in the solar system is where we find it to be. A faster growth would have lessened the chance for Earth to pick up its water, while a slower growth would have provided Earth with more water than we have now.

But these are merely theoretical arguments. We do not know how important was the role of drifting pebbles for the watering of the planetesimals picked up by Earth or Mars. In addition, there is a competing idea, which may also work. I dare call this idea my own, and I have worked on plans to run numerical simulations to evaluate its performance in detail. In Section 6.3, I described how the leftover planetesimals were cleared away from the region where Jupiter and Saturn grew due to the gravity of these planets. Many of the smaller bodies, whose diameters were just a few kilometres, would be dumped into the region just inside Jupiter's orbit because of the gas drag that they experienced.

As a consequence, the primordial planetesimals of this region — possibly almost devoid of water — were accompanied by icy colleagues from the localities beyond the snow line. Since these two groups had similar orbits, the collisions between them occurred at low speeds, so that the ice was retained. Perhaps, this was the way that Ceres and other bodies in the outer part of the asteroid belt acquired their water.

How, then, did it happen that the watered planetesimals from the outer asteroid belt were able to collide with planetary embryos in Earth's vicinity? To begin with, these two categories were far from each other, and the orbits did not allow any close encounters. One explanation is that these orbits got perturbed by Jupiter due to resonances between the revolution periods (see Section 2.2). However, this initially concerned only a minority of the bodies, which happened to have the necessary periods, and the rest must have been thrown into resonances by collisions and close encounters. A faster process would be preferable, and I think in particular of the possibility that the break down of the system of giant planets, predicted by the Nice Model, occurred very early (see Section 5.5). In this case, Jupiter's migration, even though quite small, would have wreaked dynamical havoc into the whole region somewhat closer to the Sun. Planetesimals from different zones would have been able to mix efficiently with each other.

It may be worth noting that all the embryos of Earth and Mars would also experience an invasion of icy planetesimals from the region beyond Neptune as a consequence of Neptune's migration. However, as previously mentioned, the amount of water that would thus have been swept up would have been relatively small and probably would not have competed with the amount coming from the outer asteroid belt.

I have saved the issue that is possibly the most interesting for the end. This concerns the abundance of deuterium in the water. Any acceptable theory about the origin of Earth's water must be compatible with the measured ratio between normal hydrogen (H) and heavy hydrogen or deuterium (D) in the water of the oceans. This amounts to 6420, i.e. there is one D atom in 6420 H atoms. It has been found that the water of carbonaceous chondrites — the meteorites from the outer asteroid belt — has an average value in good agreement with this oceanic value. Thus, the idea of the water originating from there stands the necessary test. But the comets exhibit discordant deuterium abundances.

The values of different comets in fact scatter over a wide interval. The lowest deuterium abundance has been measured in comet Hartley 2 (Section 1.5) and agrees with the oceanic value. The other extreme value comes from the Rosetta comet Churyumov–Gerasimenko and is three times as high. In between, there are several well-known comets, for instance Halley and Hale–Bopp, whose deuterium abundance is twice as high as that of the oceans. The reason for this scatter is thought to be that different comets were formed at different distances from the Sun and, therefore, at different temperatures — the further away, the higher the deuterium abundance.

Since all these comets come from the extensive region beyond the giant planets, it appears that the interval of deuterium abundances reflects the temperature differences within this region. The terrestrial value would then be characteristic of the region just inside, where the giant planets were formed. If so, and if the mechanism of migrating icy planetesimals was operating, it is entirely natural that the watering of planetesimals in the outer part of the asteroid belt yielded the "correct" deuterium abundance. On the other hand, the mechanism of drifting pebbles may have a problem. These pebbles would possibly originate from too large distances and carry too high deuterium abundances.

7.2. The Origin of Life

Living organisms have existed on the Earth since the planet was young. There is unfortunately no full consensus around the oldest proposed forms of life, but we can certainly claim the existence of certified finds with an age of 3.5 billion years. These are microfossiles and fossilised stromatolites (see Fig. 7.1). Stromatolites are layered colonies of cyanobacteria, and some are still living at certain places. Biogenic graphite and stromatolites have also been reported from the oldest sedimentary rocks, which have an age of 3.77 billion years. These are exposed in the Isua region of southwestern Greenland.

Fig. 7.1. Stromatolites in Shark Bay, Western Australia. Credit: Paul Harrison. License: GNU Free Documentation License.

Even older microfossiles (4.28 billion years) have been reported from metamorphic rocks of partly sedimentary origin in Québec.

All these life forms were water-living — the oldest land-living species are bacteria that thrived 3.22 billion years ago. Water is generally regarded as a *sine qua non* for the evolution of life as we know it. As seen in the preceding section, water arrived on Earth at least 4.45 billion years ago. According to some reports, there is evidence of liquid water on Earth as early as 4.40 billion years ago. The evidence is built on analysis of the mineral zircon (zirconium silicate; $ZrSiO_4$) with this extreme age.

We may ask three questions about the origin of life: when, where and how? As we have just seen, the first question can be answered by "very early". But the big unsolved problem is "how", and another interesting problem is "where". Since the DNA molecule is the cornerstone of life, the issue is about how and where this first arose. How this construction happened is not a question for astronomers, but we do take an interest in the building blocks and their origin. These are usually called prebiotic molecules, and we primarily think of amino acids.

If we suppose that life arose only, or at least for the first time, on Earth, the issue is how to explain the presence of amino acids on the surface of the very young planet Earth. There are in principle three possibilities. The first is that amino acids were part of the material that Earth was formed from. The second is that amino acids arrived somewhat later thanks to the continued planetesimal bombardment. If none of these works, one can also imagine as a third option that chemical reactions in the atmosphere of newborn Earth created amino acids.

A long time ago, at the time when the knowledge about the role of collisions in the history of the solar system was not yet established, the focus was mainly on the third possibility. In the early 1950s, a famous series of experiments was carried out in the USA by the Nobel laureate Harold Urey and his student Stanley Miller. We still refer to the Miller–Urey experiment (see Fig. 7.2). In brief, this consisted of shooting lightnings through a vessel with a mixture of gases containing carbon and nitrogen. The presence of

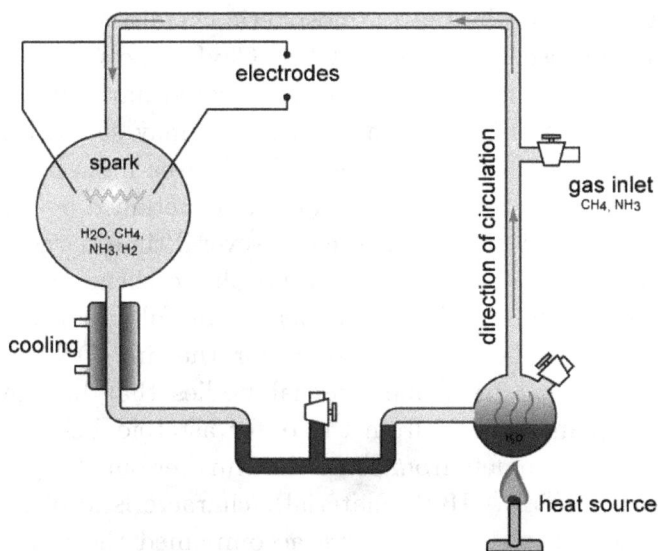

Fig. 7.2. Sketch of the setup of the Miller–Urey experiment. Credit: Hebrew Wikipedia/Carny. License: Creative Commons Attribution 2.5 Generic.

methane (CH_4) was very important. The experiment confirmed the idea that the energy of the lightnings might cause chemical reactions that produced amino acids. This result was greeted with enthusiasm and made Miller as famous as his tutor.

But there is a snag. The enthusiasm rested on the prevailing opinion of that time, according to which the early atmosphere of Earth was chemically reducing with a large content of methane. Thus, the Miller–Urey experiment provided an important argument supporting the belief that Earth's amino acids were formed chemically with no significant external influence. This belief persisted for a long time. Eventually, however, some scepticism emerged from a change of opinion about the early atmosphere and its state of oxidation. More and more experts spoke in favour of an oxidising atmosphere, where the carbon mostly appeared in the shape of carbon dioxide (CO_2) and the methane abundance was very low. In such a situation, the production of amino acids would almost vanish.

One can estimate the amount of organic material (note: organic in the sense of organic chemistry, not biologic material) that Earth would have received by both atmospheric chemistry and impacting bodies. In the case of chemistry, Chris Chyba and Carl Sagan have estimated that 1 billion tonnes per year would have been produced under the most favourable conditions. This may sound impressive, but the potential of the impacting bodies is even higher. A tentative estimate is that the total amount of organics retained by Earth from impacts around 4.4 billion years ago is several times larger than the maximum amount obtained from atmospheric chemistry during 50 million years, assuming the atmosphere to be fully reducing.

Thus, I could bet that the basis for the first DNA molecules came to Earth *via* the same celestial bodies that brought water. These were planetesimals from the outer asteroid belt, which had been watered by comets from somewhat further out. Large amounts of organic material (CHON material), characteristic of all comets (see Section 1.4), would then have accompanied the water. We do not know if the chemical or celestial organic material is best suited as a breeding ground for DNA, but one amino acid called glycine was discovered almost 50 years ago in the carbonaceous Murchison meteorite, and glycine was also identified with certainty in the outflowing material from comet Churyumov–Gerasimenko with the aid of the Rosetta probe (see Fig. 7.3).

When we consider the early planetesimal bombardment of the Earth as a source of prebiotic substances, we should also consider its negative influence. At the time when life had emerged, it may also have been exterminated by the largest impacts. This inspires questions about whether life was born, killed and resurrected time after time, and how long it took before it had the chance to survive and evolve.

Let us take a closer look at what could have happened. We assume that Earth had an ocean more than 4 billion years ago — perhaps as much as 4.4 billion. In this ocean, microscopic organisms were thriving. The first of these assumptions is very reasonable, and the second is highly uncertain but only serves as an illustration. It is beyond doubt that, at the time in question, Earth was struck by large planetesimals. Some may have had diameters exceeding 200 km, so let us imagine such an impact and watch its consequences.

Fig. 7.3. A specimen of the Murchison meteorite, exhibited at the National Museum of Natural History (Washington). Credit: Wikipedia/Basilicofresco. License: Creative Commons Attribution-Share Alike 3.0 Unported.

The enormous explosion punctuated Earth's crust and vaporised so much bedrock that the planet was surrounded by an extra atmosphere consisting of gaseous metals and silicates. This had a temperature of several thousand degrees and grilled the ground and the ocean with its radiation. As a consequence, the whole ocean boiled away during a few weeks' time, and the already very massive atmosphere was enriched by a huge quantity of water vapour. Thereafter, little by little, all came back to normal. If observers had been watching from a different place, these would have seen Earth glowing; the radiation chilled the planet until the glow waned away. During this time, stones and water droplets were regenerated. A hard rain fell for a thousand years around the globe. The result was not a downpour of 50 millimetres but 3,000 metres — the average ocean depth.

Can one imagine anything worse for living organisms? No, Earth would certainly have been sterilised, and this is the previously-mentioned extermination. Of course, the question is did this really

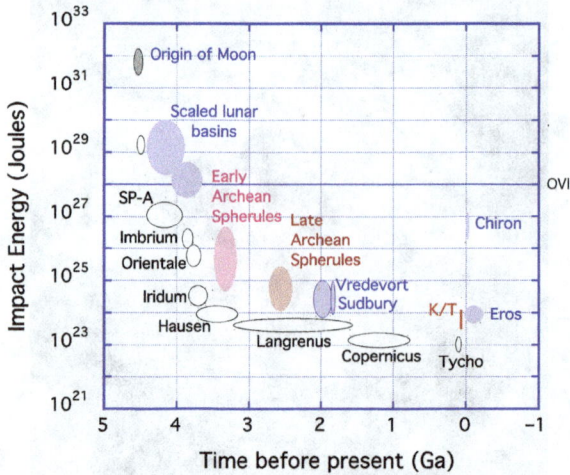

Fig. 7.4. Energies involved in the largest impact events for the Earth (shaded areas) and the Moon (open areas), plotted versus time before present. The ranges of uncertainty are given by the widths of the areas. The energy limit for an ocean-vaporising impact (OVI) is shown by a horizontal line. Original figure courtesy: K. J. Zahnle.

happen and, if so, when did it last happen? An answer is glimpsed from the statistics of large crashes into the Moon during the Late Heavy Bombardment (see Section 5.5). When we know the extent of the largest lunar impacts, we can estimate the extent of the corresponding crashes into Earth by allowing for the larger cross-section of Earth and the addition of kinetic energy due to Earth's larger gravity (see Fig. 7.4).

As a result, at least one ocean vaporising impact of the type described should have occurred at the time of this bombardment — i.e. about 4 billion years ago. Is it just by coincidence that this is just before the oldest evidence of some kind of evolved life on Earth in the form of stromatolites, or can we speculate about a causal relation? I personally find such speculation quite appealing and also well founded. It can be formulated as follows.

There are signs — though unconfirmed — that life existed on the Earth long before the end of the sterilising impacts. If life did not come from somewhere else, this means that life regenerated quickly — perhaps time after time — after having been exterminated.

Fig. 7.5. The Martian meteorite ALH 84001. Credit: NASA.

The appearance of life thus seems very likely if only the right prerequisites are present. This offers important support for the idea of extraterrestrial life. After the last sterilising impact, life comes back for good and gets established. But in the beginning, the life forms have to struggle against somewhat smaller but still devastating impacts, and this may have made an imprint on the genetic heritage that all later living organisms carry.

The latter idea means that certain impacts exterminated almost all the existing life forms. The only survivors were those that were best fitted to deal with the extreme conditions that followed upon these impacts. According to Kevin Zahnle and Norman Sleep, this means the most tolerant of heat and salt, i.e. thermophile and halophile bacteria. Life on Earth thus had to pass this filter, and Zahnle and Sleep have speculated that this may be the reason why our genetic code leads back to organisms thriving at hot volcanic gas emissions near ocean ridges. Perhaps, the case is not that this was the site where life originated, but that this was the only place where it could survive.

At last, I want to touch the classical issue of life on Mars. A little more than 20 years ago, a sensational news item was released from NASA. Some scientists claimed to have found fossil microorganisms in the Antarctic meteorite ALH84001 (see Fig. 7.5), which certainly came from Mars, representing bedrock aged 4.1 billion years. Their results have been debated and criticised, and currently, the majority opinion is that the claim was incorrect. But whatever the conclusion about this may be, I think I see good indications that an early origin of life was as likely on Mars as on Earth.

In fact, around the north pole of Mars, there is a vast region covering 40% of the planet's total area, which is situated at a level 4–5 km below the rest of the surface. Its Latin name is *Vastitas Borealis*. This may likely be the result of a giant impact in the early history of Mars. The most probable timing seems to be about 4.4 billion years ago. At that time, Mars had acquired its water, amounting to about one per mille of the planetary mass, the same as on Earth. The giant impact triggered the outgassing of an enormous amount of water vapour, which gave Mars a thick atmosphere. This was gradually lost by photochemical erosion, but in the meantime provided a good background for a life-friendly climate through its greenhouse effect. Since Mars should have acquired an amount of amino acids as important as that of the Earth by the early planetesimal bombardment, it probably offered good conditions for the origin of life.

It is related to the possibility of transporting living organisms between Mars and Earth. Once — I think it was in 1997 — I went to Gothenburg for a large conference on astrobiology in the wake of ALH84001. As I strolled past the poster exhibition, I got very interested by one, which dealt with transport of living organisms from planet to planet by means of meteorites. Standing next to the poster was an elderly gentleman, with whom I at once started to talk. This was Curt Mileikowsky, former Vice Managing Director of ASEA (now part of ABB) and Managing Director of Saab automobile company. I had heard a little of his background and was fascinated that he had started to take interest in the same things as myself.

From this meeting emerged a collaboration, which is among the most important and the most fun that I have ever had. Curt's background was in nuclear physics, and he had held a professorship at the Royal Polytechnic in Stockholm when he was young. His goal was to find out if the most radiation-resistant bacteria known to exist (e.g. *Deinococcus Radiodurans*) would possibly survive a journey between Mars and Earth, embedded into metre-sized meteoroids. He had worked with NASA scientists to measure how large a radiation dose the bacteria could survive. It then remained to compare this to the expected cosmic radiation dose during a trip from Mars to Earth or the reverse. Other effects were important, too, and some of them, like the shock effects of the impact that sent out the meteoroids, had already been investigated. I contributed with a calculation of the internal temperatures of these small bodies during their trips. I also put Curt in contact with other experts, who, for instance, could estimate how much time the trip would take. Eventually, we published a paper that is more cited than all the others that I produced during my whole career.

In this paper, we established that the transport between the two planets may work without major problems. It is easier from Mars to Earth, but the trip from Earth to Mars is also possible. For sure, the survival rate of the organisms is very low, but if there are large enough numbers of them from the start, it is enough to guarantee that either planet can be fertilised from the other. What does this imply? Since the young planets experienced a constant bombardment, which sent out lots of meteoroids, we draw an essential conclusion. In case life first arose on Mars, it certainly would make its way to Earth, and the inverse is true as well.

Are we then able to say on which planet life indeed arose? No, I don't think so. Ultimately, we may all be Martians. Is it then possible to say for sure that the origin of life was a simple matter? This would certainly be premature, but it is possible that Mars offered a more protected environment than Earth. In case life on Earth was not easily resurrected after extermination, it might be thanks to re-fertilisation from Mars that it finally had the chance to survive for good. If, one day, living organisms are indeed found on Mars, I do look forward to learning from these what really happened in the ancient times, when the solar system was still young.

7.3. Evolution and the Comets

The concept of security has an important place in our life stance and worldview. Regardless of whether we are traditionalists or revolutionaries, we want to live on solid ground in a world that stays stable for eternal times. But at the same time, we also have curiosity and a wish to understand, and from this there may emerge a worldview with troublesome properties. This is particularly noticeable in our time but was absent in Antiquity.

In the worldview of Antiquity, which remained prevalent in Europe during the Middle Ages, the world where we live — the equivalent of what we now call planet Earth — was safely anchored at the centre of a series of crystal spheres. The Sun, the Moon and planets were deported to these and moved there according to more or less intricate patterns. Comets were considered to belong to our atmosphere, and asteroids were not known at all. There was no place for collisions or other natural hazards coming from outside, unless by divine intervention.

But, of course, this does not mean that the world was a paradise. Everybody knew that terrible things happened, like wars and slavery, starvation and plagues. It was generally considered likely that these could be foreboded by omens in the sky, and in this context comets had a given role to play. Their behaviour seemed inexplicable, and it was natural to regard them as celestial harbingers. Such was the case also in China, where the comets warned of both upcoming disasters and the consequences of committed crimes or sins. If the order of society had been broken due to the Emperor's negligence, this would be seen on the sky as abnormal phenomena like, for instance, comets.

In the cultural sphere of Europe, the cometary omens were almost always evil. I can only think of one counterexample, namely, that the star of Bethlehem has sometimes been dressed in cometary attire. In the Scrovegni chapel of Padua, one can see the famous fresco by Giotto di Bondone from the first years of the 14th century, depicting the adoration of the Magi. This presents a sky decorated by a comet. As I mentioned in Section 1.2, this may have been inspired by a real comet that Giotto had seen, namely, Halley's comet.

Among the European peasant population, the comets mostly inspired worry and concern, but it is only in modern times that comet scare has blossomed out and taken frightening proportions. On some occasions, science has been invoked, like in the case of Halley's comet in 1910, when spectroscopic studies confirmed that the comet contains poisonous gases. This led to speculations about perilous consequences of Earth's passage through the comet tail, and many people were scared — completely needlessly. Worse consequences arose in the case of comet Kohoutek in 1973 (see Section 4.2). The American sect Children of God spread their own horror vision of the comet foreboding huge disaster, and the members fled to remote places, awaiting the world's demise. Tragic events also unfolded in the wake of comet Hale–Bopp. Thirty-nine members of the sect Heaven's Gate committed common ritual suicide near San Diego in March 1997, in an attempt to reach an alleged UFO travelling in the comet's vicinity.

I have said enough of what people have believed that comets can do. Now I come to what they can actually do — and may really have done — to influence the biological evolution on our planet. This concerns the largest impacts that the Earth has experienced, since evolution took off in earnest about 540 million years ago. It does not matter if the projectiles were comets or asteroids. The physical or chemical properties of the material have very little significance — the effects of the impact are primarily determined by the mass and velocity of the projectile, i.e. the kinetic energy involved, and how the ground at the impact site is structured.

Yet, it is easy to realise that an important fraction of the largest impacts may have been caused by comets. It may suffice to observe today's reality. There are one thousand Near Earth Asteroids with sizes of one or a few kilometres, but only a few tens of short-period comets. The rate of passages of long-period comets is also fairly low, and thus, it is generally agreed that the short-term impact hazard is mainly due to the asteroids. However, the situation is different when we consider the largest objects and much longer time scales. The largest Near Earth Asteroids are easily matched by the corresponding comets. The largest diameter among asteroids which cross the Earth's orbit during each revolution (the Apollo asteroids) is 8.5 km, and the record holder is named Sisyphus. We may compare this to the parent comet of the Perseid meteors, Swift–Tuttle (see Fig. 7.6), whose nucleus is thought to have a diameter of 26 km.

Fig. 7.6. Comet Swift–Tuttle, parent of the Perseid meteor stream. Credit: NASA.

Moreover, at the current time, Sisyphus cannot approach Earth within 15 million km, while Swift–Tuttle may in principle come close to collision at any time. We are partly protected by the relatively long period of this comet (130 years), which implies that the comet is almost always very far away. It passes close to Earth's orbit once per revolution, and the risk for Earth to be there at the wrong time is very small. In addition, the comet orbit turns slowly, and the present — relatively hazardous — configuration is rare in the long run. A collision thus requires a very unlikely combination of different circumstances, and the risk is nearly vanishing. The result is that a comet impact by such a large projectile is only expected to happen once in a 100 million years.

In fact, a similar estimate can be made for asteroidal projectiles, too. All we can say is that impacts causing 100 km sized craters are exceedingly rare and that neither comets nor asteroids can be excluded as projectiles. Since the traces of these impacts are gradually wiped out over time, in spite of the large size, we have to face the fact that Nature provides us with very limited evidence. Safe conclusions are therefore very difficult to reach. I will describe a few examples.

In Section 5.3, I mentioned the large Chicxulub crater, which was formed 65 million years ago at the turn of the Cretaceous and

Tertiary epochs (the K/T boundary) and is associated with a mass extinction, which occurred just at that time and — among others — struck the dinosaurs. There are two more craters of similar size, i.e. roughly 100 km in diameter, from the last 250 million years, while the evidence for older craters of comparable size is unclear due to severe erosional modification. There were also two more mass extinctions of the same importance as the K/T extinction, which occurred during the same period of time. The two craters are Popigai in northern Siberia, 90 km in diameter, and Manicouagan, 85 km in diameter, situated in the Canadian province Québec (see Fig. 7.7).

The age of the Manicouagan crater is 214 million years, which is close to, but not coinciding with, a large mass extinction at the turn of the Triassic and the Jurassic 205 million years ago. Due to the time offset, this crater cannot be considered a likely reason for the extinction. Nonetheless, shocked quartz (see Section 5.2) has been found in the boundary layer, indicating that another large impact

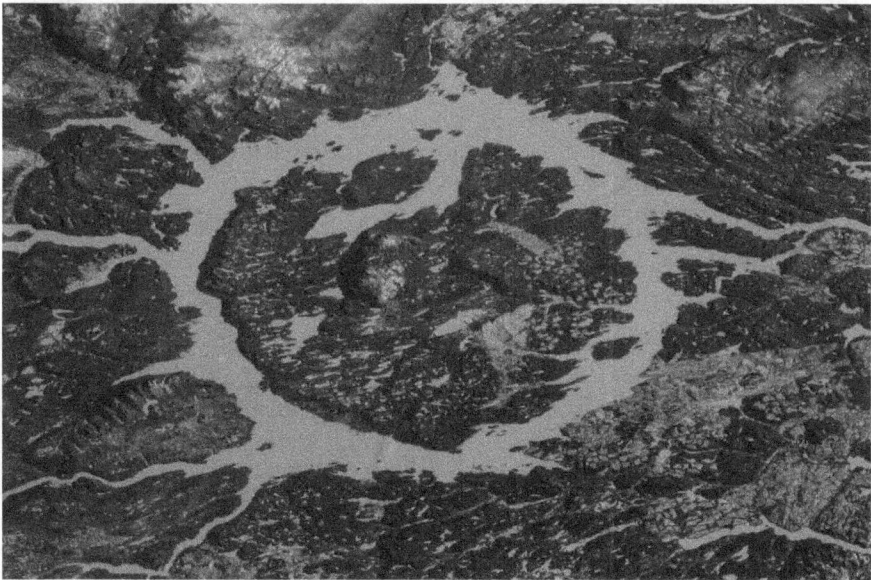

Fig. 7.7. The Manicouagan impact crater imaged from the International Space Station. The picture features an ice-covered annular lake surrounding an inner island plateau. Credit: NASA.

did occur and might have played a significant role. However, no corresponding crater has been found. For Popigai, the situation seems clearer. Its age is 36 million years. About 25 years ago, it was clarified that another crater of considerable size (40 km) was almost the same age. It is situated in Chesapeake Bay at the US east coast. Within likely dating errors, there is even a chance that these two craters were actually formed at precisely the same time as parts of the same impact event.

Concurrently, there was a mass extinction of considerably smaller extent than the two just mentioned. It is undisputed and corresponds to the boundary layer between the Eocene and the Oligocene. Shocked quartz has been found here, too, which is a good indication of one or more impacts. Perhaps, as indicated, one large body crashed into Earth but broke apart immediately before the collision, thus causing the two detected craters. Otherwise, we have an indication that Earth, the Moon and the planets were ravaged by a very fierce shower of projectiles at the time in question. In both cases, a cometary connection seems likely, but as yet, this is purely a matter of speculation.

In the 1980s, comets caught attention for a special reason. Some researchers claimed to see a periodic signal in the time series of mass extinctions. This purported signal was controversial from the very start, and it has not survived the subsequent debate. However, more than 30 years ago, it still attracted great interest. An issue arose as to which natural process might cause mass extinctions of a regular periodicity, and the only reasonable suggestion was comet impacts. In this case, it would certainly be a question of comet showers from the Oort cloud (see Section 4.1).

In the case at hand, the period is about 30 million years. I recall two suggestions as to how comet showers could appear with such a periodicity. One of them invokes the orbital motion of the Sun in the Galaxy. While the Sun completes an orbit around the centre in about 240 million years, it also oscillates up and down through the central plane. The passages through this plane occur at intervals of somewhat more than 30 million years. Both a majority of the most massive stars and the supermassive molecular clouds where star formation is initiated (see Section 3.4) are concentrated toward the central plane. The idea was that the Sun would therefore run the risk of close encounters with such objects during its passages through the plane, and as a consequence, the entire Oort cloud would be shaken by the gravity of the object, thus causing a comet shower.

The other suggestion involved a daring hypothesis. The Sun would be accompanied by an as yet undiscovered dwarf star or giant planet, unofficially named Nemesis. It would be orbiting at the very large distances of the Oort cloud. The idea meant that Nemesis would have a strongly elliptic orbit with a period of about 30 million years, so that it periodically passes through the tightly populated inner core, thus causing comet showers. In retrospect, I note that both suggestions are vulnerable to devastating criticism, and I only mention them as examples of scientific wanderings. Above all, their logical basis has now disappeared, since very few scientists still persevere in the claim of periodicity in the mass extinctions.

Let us finally return to the fundamental issue: Are the mass extinctions on Earth caused by gigantic cosmic impacts? There are different opinions about this, and I can only present my own. As a matter of fact, plant and animal species die out more or less continuously, so there is a background from which the mass extinctions have to stand out. Thus, a limit has to be passed in order for us to say that something special has occurred. This limits the number of episodes that are deemed worthy of discussion. But even if we consider only the clearest cases, there remain some that show an evident connection to impacts. There are likely several reasons for mass extinctions, but the impacts definitely ought to be one of these.

As to the biological evolution, this may principally have proceeded quietly and gradually in accordance with Darwin's hypothesis of natural selection. But now and then, all that this evolution has accomplished is upset when mass extinctions arrive. The old world is ravaged, but like a phoenix, new life emerges from the ashes of the old as the survivors enjoy leeway, dominate the world and are able to evolve in peace and quiet until it is time for them to be struck. This seems to be the fact: evolution and disasters go hand in hand. Nothing lasts forever, and all has its allotted time.

In this spectacle, the celestial bodies play an important role. They can be viewed as arbiters of life on Earth, who intervene very seldom but very thoroughly. In a somewhat loaded wording, we can say that comets and asteroids have brought water, life and death to our planet, and without mercy, they continue to rule over life and death.

Index

www.ingramcontent.com/pod-product-compliance
Lightning Source LLC
Chambersburg PA
CBHW061246220326
41599CB00028B/5549